Emerging Advancements in Agricultural Sector

A Progressive Approach

Emerging Advancements in Agricultural Sector

A Progressive Approach

Edited by

Manisha Vohra

CWP

Central West Publishing

Disclaimer
Every effort has been made by the publisher, editors and authors while preparing this book, however, no warranties are made regarding the accuracy and completeness of the content. The publisher, editors and authors disclaim without any limitation all warranties as well as any implied warranties about sales, along with fitness of the content for a particular purpose. Citation of any website and other information sources does not mean any endorsement from the publisher, editors and authors. For ascertaining the suitability of the contents contained herein for a particular lab or commercial use, consultation with the subject expert is needed. In addition, while using the information and methods contained herein, the practitioners and researchers need to be mindful for their own safety, along with the safety of others, including the professional parties and premises for whom they have professional responsibility. To the fullest extent of law, the publisher, editors and authors are not liable in all circumstances (special, incidental, and consequential) for any injury and/or damage to persons and property, along with any potential loss of profit and other commercial damages due to the use of any methods, products, guidelines, procedures contained in the material herein.

NATIONAL LIBRARY OF AUSTRALIA

A catalogue record for this book is available from the National Library of Australia

ISBN (print): 978-1-922617-37-8

About the Book

Agriculture is one of the utmost vital and utmost essential sector of the society. It is utmost vital and essential as it provides us with food without which it is not possible to survive. Each sector undergoes advancements for its betterment. Similarly, agriculture sector is also moving in this direction.

This book titled Emerging Advancements in Agricultural Sector: A Progressive Approach is a collection of book chapters explaining different aspects related to agriculture and the advancements that are emerging in it. There are various emerging advancements in the sector of agriculture which are seeming to be helpful for this sector. The advancements are coming across as a progressive approach as they can be useful for the betterment of this sector and even help the farmers in their different agricultural work.

Chapter 1:
Different sectors such as education, healthcare, etc. are moving towards advancements in recent times. Likewise, agriculture sector is also moving towards advancements. Agriculture sector is very important for getting food. Agriculture sector provides with various kind of food production. Agriculture depends on various factors like soil, climate, etc. There are many things to look after in agriculture like irrigation, requirement of compost etc. In this book chapter, a brief discussion on agriculture and its insight is done. Also, a brief discussion is done, and related explanation is provided on agriculture moving towards the path of advancements, what is the need for it and its importance.

Chapter 2:
People are to a great extent dependent on agriculture for getting food. In this work to have innovative irrigation for agriculture, two new models for irrigation purpose have been described. Mostly for irrigation purpose, electricity, petrol and is required and sometimes unavailability of electric supply and increment in petrol and diesel price causes the irrigation to suffer. To solve these issues, two model are designed by using mechanical concepts that will be useful to farmers for lifting water for irrigation purpose without any external sources. The use of these designed models is very easy for farmers. The designed model are environment friendly with less cost. Only purchase

and installation cost initially is required for this purpose. Repetitive expenditure for irrigation purpose is restricted by these models.

Chapter 3:
In agriculture when it comes to growing any type of crops, there are various factors in it which need to be looked after and taken care of like soil condition, climate, compost availability, etc. However, one of the most important factors is irrigation which needs to be carried out on a regular basis, throughout the period when the crop is being grown. To efficiently carry out irrigation in a way that proves to be helpful and eases the work of farmers, various advanced irrigation systems are being talked about, discussed, proposed and even implemented by various experts, professionals, etc. A literature review is carried out in this chapter to understand these advanced irrigation systems in agriculture. A comparison of the reviewed systems is also done in the chapter.

Chapter 4:
Agriculture forms a significant part of every country's economic growth. It has been a field of deep research in the past and it continues to be. Many researchers are carrying out research in the field of agriculture in improving the betterment of agricultural production. To predict the yield of agricultural crops machine learning models can be used in prediction based on parameters like weather data, nature of the soil at the crop cultivation environment. In this chapter the prediction models like random forest and decision trees are used to predict the yield of crops in advance based on the weather, soil and historic crop yield data, which will help the farmers in identifying the suitable crop for cultivation. The performance of the prediction models was evaluated on the crop yield data of maize collected for different seasons and the results inferred that random forest model outperforms decision tree models in predicting the crop yield.

Chapter 5:
Plant leaf disease is a major problem in the agricultural sector. In various seasons, several plant leaf diseases appear. A deep learning approach, which can help farmers understand the plant leaf disease and identify them can be used. In the area of research that involves the feature extraction for plant leaf disease using image processing techniques can help farmers make optimal decisions quickly and accurately. A study has been carried out in this chapter and on its basis

the authors have proposed a system where they have used image pre-processing and image augmentation techniques to obtain a better image, which allow them to process better results for analysis. In this study, three types of leaves take into consideration, namely bell pepper, potato and tomato leaves. Classification is performed in the proposed system for classifying plants leaf disease of bell pepper, potato, and tomato leaves. The system is divided into two stages. In the first stage, classifier, a median filter and image augmentation are used and the algorithm is trained and second stage classifier, the extract plant leaf image output is performed using ResNet50. So, with two step classification approaches and a deep learning approach to respective plant leaves for observing the diseases, 98% accuracy for detecting and classifying the leaf disease.

Chapter 6:
Machine Learning is gaining more popularity nowadays due to the increase in the generation of a huge amount of data every day by a single individual, lots of mathematics bases are available for calculations and many new technologies like big data and high computing processors are making it possible to process those data. So, machine learning techniques can also be implemented along with agro-technologies to improve productivity in agriculture. The chapter includes an overall review of how these algorithms have been applied to various divisions in agriculture management. The chapter talks about crop management, soil management, water management, livestock management and other such things related to agriculture. The review in this chapter will help in understanding how different machine learning algorithms can help farmers in making the best decision for their crops and lands. Also, it can help the researchers to identify how and which applied algorithms can be optimized by tuning their parameters.

Chapter 7:
The use of Artificial Intelligence (AI) is constantly increasing in the recent past. AI is being used in many different sectors like industrial sector, healthcare sector, etc. due to the advantages it provides. In agricultural sector also, AI is finding and making its place. Agriculture is a sector which is undoubtedly one of the most important and necessary sectors. The application of AI in agriculture can be of great help. This book chapter presents a systematic study of AI application in ag-

riculture sector based on literature. A real world AI application example in agriculture is also discussed. Along with it, the advantages of AI
in agriculture are also discussed in this chapter.

Table of Contents

1

Moving Towards the Path of Advancement in Agriculture: Need and Importance

Manisha Vohra
Independent Researcher, India

Abstract: Different sectors such as education, healthcare, etc. are moving towards advancements in recent times. Likewise, the agriculture sector is also moving towards advancements. The agriculture sector is very important for getting food. Agriculture depends on various factors like soil, climate, etc. There are many things to look after in agriculture like irrigation, the requirement of compost, etc. In this book chapter, a brief discussion on agriculture and its insight is done. Also, a brief discussion is done on agriculture moving towards the path of advancements, what is the need for it and its importance, and along with it, a related explanation regarding the same is provided in the chapter.

1. Introduction

In today's world, everything is advancing, right from the healthcare sector where through video calls, doctors can be consulted in urgent situations to the education sector where online classes can be conducted and similarly in various other sectors, different advancements are witnessed. The agricultural sector also is witnessing advancements which will be discussed and understood with the help of some examples further on in the chapter.

The agricultural sector is the backbone sector of the society as it looks after food production needs. The agriculture sector largely contributes as a very major source of food production. Just like oxygen, food is also very essential for the existence of humans.

Since agriculture serves as a food-producing source, it can be correctly stated that agriculture is a boon to the entire mankind. Agriculture has various benefits and it also provides employment to several people who undertake farming as their occupation.

Along with it, agriculture also helps in improving the economy of a country. Be it highly consumed foods like rice, wheat, vegetables, fruits, etc. to things like cotton, which is used in the medical sector and in other sectors like industrial, etc. for making clothes, bags, etc. to jute which is used for making bags, etc. agriculture provides us with this all. Hence agriculture is very beneficial.

Besides, agriculture also provides herbs. There are different types of herbs that we can get from agriculture like consumable herbs, medicinal, etc. So altogether, agriculture is very vital, along with being beneficial.

2. Insight into Agriculture

To grow anything in the agricultural field involves a lot of effort and hard work. The farmers really work very hard. Right from making the agricultural field or land ready for sowing seeds for growing crops to harvesting the grown crops is an intricate process.

In agricultural work, some of the tasks include the land survey and its soil testing where crops are decided to be grown. The regional climatic conditions and weather-related information of the previous years are also taken into consideration and then depending on these factors and also some other factors as per the requirement, a specific type of crop is zeroed in for growing.

Water availability and electricity are also key factors for agriculture. Water for irrigation and electricity for the pump to be operated are needed. Thus, agriculture depends on various factors and all of them need to be looked after and taken care of.

Providing an appropriate amount of water on a timely basis i.e., the irrigation process is one of the most essential requirements for the proper growth of all types of crops undertaken for growth.

Crops can take a few weeks to many months to grow. It all depends on what is exactly being grown. So, during this entire time of crop growth, which begins from the time the seeds are sowed in the soil, till the time crops are completely grown and can be harvested, irrigation has to be carried out as per the requirement.

Irrigation cannot be avoided. The crops will not be able to survive without irrigation. The farmers have to regularly monitor the crops and as per the requirement of the crops, irrigation needs to be carried out in the field.

Depending on the feasibility of the irrigation technique and its associated resource requirements, the best possible irrigation technique is practiced by each farmer as per their suitability and convenience. There are different types of irrigation techniques present. Some of the famous irrigation techniques practiced are enlisted below:

- Drip Irrigation
- Sprinkler Irrigation
- Surface Irrigation
- Center Pivot Irrigation

Besides irrigation, other factors like protection against insects through pesticides, proper use of compost, etc. are necessary for the growth of crops.

Unfortunately, despite of a lot of efforts put by farmers and great farming practices being put into practice by them some crops get spoilt sometimes due to different reasons, for example, floods, and drought.

3. Need for and Importance of Advancement in Agriculture

In today's world, a lot of technologies are present. They are being utilized across various sectors with absolute success. With the presence and application of different technologies, there is an increase in advancements in most of the sectors, be it the industrial sector or educational sector, etc.

Similarly, in the agricultural sector as well, there are different emerging advancements due to the presence and application of different technologies. Various technology-based systems are being utilized and suggested by experts to be used in the agriculture sector. Hence, the agricultural sector is also rightly moving toward the path of advancement.

These advancements include the utilization of various systems like automatic irrigation and monitoring of crops with the help of internet of things (IoT) based systems, artificial neural network (ANN) based controller for irrigation, neural network-based plant disease identification and classification, etc. which seem to be beneficial for this sector. Let us understand these with the help of examples.

Example 1:

Consider the example of the article [5]. In this study, the objective of the authors is of finding out blast disease and getting a decrease in crop loss and hence ultimately get an increase in rice production in an efficient manner. Deep convolutional neural networks have been used by the authors here and image classification is done utilizing them [5]. According to the results obtained from the developed model of the authors, pests and rice diseases can be very well detected and realized including healthy plant class with the usage of a deep convolutional neural network, having the optimum accuracy of 96.50% [5].

Example 2:

Consider the example of the article [28]. Here, in this study, an artificial neural network (ANN) based controller for irrigation is proposed by the authors. The evapotranspiration model is also brought to usage. Overall, through this system, it was found that by using ANN, the authors are obtaining accurate results when they calculate the required soil moisture from input parameters, and this is very useful and helpful in carrying out irrigation [28].

Every irrigation technique involves human efforts. To ease things for the farmers, automatic irrigation will be a great help.

Let us understand this briefly as this explains both, the need as well as the importance of moving towards the path of advancement in agriculture. Continuing with the same example of irrigation, it is known that it requires time and effort.

Irrigation is to be carried out on a time-to-time basis as per the need. So throughout the period of time when the crops are growing, starting from the time seeds are sowed, till the time crops are completely grown and ready for harvest, irrigation is to be carried out on a time-to-time basis as per the need. It requires effort and time. A lot of time and effort of the farmers are spent in carrying out irrigation.

In agriculture, the farmer has a lot of work like irrigation, checking for the need for compost, spraying pesticides, etc. If irrigation can be automated, then it will provide farmers with more time for other agricultural work, some of the time and effort of the farmers will also be saved.

Irrigation is not something that is carried out once or twice during the period the crops are being grown. Irrigation cannot be even skipped. It is very important in agriculture for the crops being grown. Hence advancements like these are needed and have importance as they can help farmers to ease out their work and save some of their time and effort. Advancements like automatic irrigation as irrigation can be carried out without anyone's help and it can save water as well which is very beneficial.

Other advancements for example neural network-based systems, etc., which help in the case of plant diseases, are also needed and have importance as instead of manually trying to detect and recognize plant diseases, different systems like the one explained in example 1 [5] can work well thus easing work and enabling to gain information regarding the diseases in the plants quickly and effectively.

4. Discussion

The examples from the literature discussed above which talk about advancements in agriculture are proposed and even tested by various experts. Even though a large-scale application of these advancements is not yet implemented worldwide but with various such systems already proposed, tested and implemented and such systems are still being proposed and tested with different variations and methods by experts, there has been a shift towards advancements in agriculture. Likewise, other advancements in agriculture, which on testing can prove to be helpful, can be useful. There are various possibilities of witnessing more advancements in agriculture and if any systems of testing can prove to be helpful then this could be useful and such advancements would be considered useful for agriculture.

5. Conclusion

A brief discussion is provided on agriculture in this book chapter. Along with it, a brief discussion and explanation of the need for and importance of agriculture moving towards the path of advancements is also presented in this chapter.

With so many advancements going on everywhere, in each sector, the agriculture sector also saw a shift. It is also moving towards advancements. Different advancements with the help of examples were discussed in the chapter.

Advancements in agriculture which prove to be useful in testing, can help to ease out the work of the farmers and be beneficial for the agriculture sector as well.

References

1. Arvindan, A. N., Keerthika, D. (2016) Experimental investigation of remote control via Android smart phone of arduino-based automated irrigation system using moisture sensor. *2016 3rd International Conference on Electrical Energy Systems (ICEES)*, pp. 168-175.
2. Nikolidakis, S. A., Kandris, D., Vergados, D. D., Douligeris, C. (2015) Energy efficient automated control of irrigation in agriculture by using wireless sensor networks, *Computer and Electronics in Agriculture*, 113, 154-163.
3. Angal, S. (2016). Raspberry pi and Arduino Based Automated Irrigation System. *International Journal of Science and Research (IJSR)*, 5(7), 1145-1148.
4. Kodali, R. K., Jain, V., Karagwal, S. (2016) IoT based smart greenhouse. 2016 IEEE Region 10 Humanitarian Technology Conference (R10-HTC), pp.1-6.
5. Bharathi, R. J. (2020) Paddy Plant Disease Identification and Classification of Image Using AlexNet Model. *The International journal of analytical and experimental modal analysis*. XII(III), pp. 1094-1098.
6. Kansara, K., Zaveri, V., Shah, S., Delwadkar, S., Jani, K. (2015). Sensor based Automated Irrigation System with IOT. *International Journal of Computer Science and Information Technologies (IJCSIT)*, 6 (6), 5331-5333.
7. Kawasaki, Y., Matsuo, S., Suzuki, K., Kanayama, Y., Kanahama, K. (2013). Root-Zone Cooling at High Air Temperatures Enhances Physiological Activities and Internal Structures of Roots in Young

Tomato Plants. Journal of the Japanese Society for Horticultural Science, 82(4), 322-327.

8. Nishina, H. (2015). Development of Speaking Plant Approach Technique for Intelligent Greenhouse. *Agriculture and Agricultural Science Procedia*, 3, 9–13.

9. Liakos, K. G., Busato, P., Moshou, D., Pearson, S., Bochtis, D. (2018). Machine Learning in Agriculture: A Review. *Sensors*, 18(8), 2674, 1-29.

10. Venugoban, K., Ramanan, A. (2014). Image classification of paddy field insect pest using gradient-based features, *International Journal of Machine Learning and Computing (IJMLC)*, 4 (1), 1-5.

11. Sardini, E., Serpelloni, M. (2011). Self-powered wireless sensor for air temperature and velocity measurements with energy harvesting capability. *IEEE Transactions on Instrumentation and Measurement*, 60(5), 1838-1844.

12. Vaishali, S., Suraj, S., Vignesh, G., Dhivya, S., Udhayakumar, S. (2017). Mobile integrated smart irrigation management and monitoring system using IOT. *International Conference on Communication and Signal Processing (ICCSP)*, 2164-2167.

13. Gutirrez, J., Francisco, J., Villa-Medina, J. F., Nieto-Garibay, A., Porta-Gándara, M. Á. (2014). Automated Irrigation System Using a Wireless Sensor Network and GPRS module, *IEEE Transactions on Instrumentation and Measurement*, 63(1), 166-176.

14. Pavithra, D. S., Srinath, M. S., (2014). GSM based Automatic Irrigation Control System for Efficient Use of Resources and Crop Planning by Using an Android Mobile. *IOSR Journal of Mechanical and Civil Engineering (IOSR-JMCE)*, 11(4), 49-55.

15. Rajalakshmi, P., Mahalakshmi, S. D., (2016). IOT based crop-field monitoring and irrigation automation, *2016 10th International Conference on Intelligent Systems and Control (ISCO)*, 1-6.

16. Pernapati, K. (2018). IoT based low cost smart irrigation system. *Second International Conference on Inventive Communication and Computational Technologies (ICICCT)*, 1312- 1315.

17. Davis, S. L., Dukes, M. D. (2010). Irrigation scheduling performance by evapotranspiration-based controllers. *Agricultural Water Management*, 98 (1), 19-28.

18. Agarwal, M., Kaliyar, R. K., Singal, G., Gupta, S. K. (2019) FCNN-LDA: A Faster Convolution Neural Network model for Leaf Disease identification on Apple's leaf dataset, *12th International Conference on Information & Communication Technology and System (ICTS)*, pp. 246-251.

19. Arun Pandian, J. A., Geetharamani, B. Annette. (2019) Data Augmentation on Plant Leaf Disease Image Dataset Using Image Manipulation and Deep Learning Techniques, *IEEE 9th International Conference on Advanced Computing (IACC)*, 2019, pp. 199-204.

20. M. Pagliai, N. Vignozzi, S. Pellegrini, (2004) Soil structure and the effect of management practices, *Soil and Tillage Research*, Vol. 79, No. 2, pp. 131-143.
21. Patil, C., Aghav, S., Sangale, S., Patil, S., Aher, J. (2021) Smart Irrigation Using Decision Tree, *Advances in Intelligent Systems and Computing*, 1245, pp. 737-744.
22. Nasiakou, A., Vavalis, M., Zimeris, D. (2016) Smart energy for smart irrigation, *Computers and Electronics in Agriculture*, 129, pp. 74-83.
23. Kumar, V., Arora, H., Harsh, Sisodia, J. (2020) ResNet-based approach for Detection and Classification of Plant Leaf Diseases, *International Conference on Electronics and Sustainable Communication Systems (ICESC)*, pp. 495-502.
24. Shruthi, U., Nagaveni, V., Raghavendra, B. K. (2019) A Review on Machine Learning Classification Techniques for Plant Disease Detection. 2019 5th International Conference on Advanced Computing & Communication Systems (ICACCS). IEEE, 2019.
25. Dubey, V. (2011) Wireless Sensor Network Based Remote Irrigation Control System and Automation Using DTMF Code, *International Conference on Communication Systems and Network Technologies (CSNT)*, 2011, pp.34 –37.
26. Dhakate, M., & Ingole, A. B. (2015, December). Diagnosis of pomegranate plant diseases using neural network. *2015 fifth national conference on computer vision, pattern recognition, image processing and graphics (NCVPRIPG)* pp. 1-4.
27. Viani, F., Bertolli, M., Salucci, M., Polo, A., (2017). Low-cost wireless monitoring and decision support for water saving in agriculture. *IEEE Sensors Journal*, 17(13), 4299-4309.
28. Raja Sekhar Reddy. G, Manujunatha. S, Sundeep Kumar K. (2013). Evapotranpiration Model Using AI Controller for automatic Irrigation system. *International Journal of Computer Trends and Technology (IJCTT)*, 4(7), 2311-2315.
29. Mohanty, S. P., Hughes, D. P., Salathé, M. (2016) Using deep learning for image-based plant disease detection. *Frontiers in plant science* 7, article 1419, 1-10.
30. S. Dhivya, Shanmugavadivu, R. (2018) Comparative Study on Classification Algorithms for Plant Leaves Disease Detection. *International Journal of Computer Trends and Technology (IJCTT)*, 60(2), 115-119.
31. Barbedo, J. G. A. (2018) Factors influencing the use of deep learning for plant disease recognition. *Biosystems engineering* 172, 84-91.
32. Yarak, K., Witayangkurn, A., Kritiyutanont, K., Arunplod, C., Shibasaki, R. (2021) Oil Palm Tree Detection and Health Classification on High-Resolution Imagery Using Deep Learning. *Agriculture*, 11(2), 183.

33. Ojha, T., Misra, S., Raghuwanshi, N. S., (2015). Wireless sensor networks for agriculture: the state-of-the-art in practice and future challenges. *Computers and Electronics in Agriculture*, 118, 66–84.

Innovative Irrigation for Agriculture

Ramesh Chandra Nayak
Department of Mechanical Engineering, Synergy Institute of
Technology, Bhubaneswar, Odisha, India

Manmatha K. Roul
Department of Mechanical Engineering, Gandhi Institute for
Technological Advancement (GITA), Bhubaneswar, Odisha, India

Saroj Kumar Sarangi
Department of Mechanical Engineering, NIT Patna, Bihar, India

Abstract: People are to a great extent dependent on agriculture for getting food. In this work to have innovative irrigation for agriculture, two new models for irrigation purpose have been described. Mostly for irrigation purpose, electricity, petrol and is required and sometimes unavailability of electric supply and increment in petrol and diesel price causes the irrigation to suffer. To solve these issues, two model are designed by using mechanical concepts that will be useful to farmers for lifting water for irrigation purpose without any external sources. The use of these designed models is very easy for farmers. The designed models are environment friendly with less cost. Only purchase and installation cost initially is required for this purpose. Repetitive expenditure for irrigation purpose is restricted by these models.

1. Introduction

A better environment in a workplace enhances productivity. At many places, till now farmers are using traditional methods for their agriculture. However, with advancements and new and improved techniques in agriculture, productivity can be enhanced. In agriculture, irrigation for crops is a major requirement. Expenditure for irrigation is also high. In different places, irrigation by farmers is carried out by petrol, kerosene and electric operated pumps. In this chapter we have presented our work which will help farmers in their irrigation. Our work has been implemented and verified in Cuttack district of state Odisha. Whenever there are cyclones, during these periods unavailability of petrol, diesel, kerosene and failure of electric supply occur which disrupts irrigation process for a long time which impacts the crops largely. To overcome this problem, we have developed a new method and designed a new model for irrigation that will help farmers for their irrigation purpose without requirement of any external sources like petrol, kerosene, etc. which will be explained in the chapter later on.

There are number of works that has been done by researchers on different irrigation methods which we will understand through literature review presented as follows.

Literature Review

The authors in paper [2] developed a new tool for enhancing productivity of agriculture; the web-based tool is intended to guide farmers in their opinion to invest in certain tools from protected areas, growing the stage of knowledge about the features of new technologies and the associated settlement. The potential diminution of inputs also offers the opportunity of studying the alleviation of environmental blows.

Authors in paper [3] found that the results specify that participation in CATP boosts the approval of climate-smart cowpea arrays, cowpea yield and income by 75, 15, and 24%, respectively, from their average levels. These optimistic happiness impacts of participation in CATPs underscore the requirement to increase capacity building activities in agricultural enhancement projects and to intend method to remove obstacles to involvement of rural farm families.

Paper [5] discusses some of the challenges that require to be addressed to make certain that smart tools and big data can assist farmers to surmount external blows. In their work, the authors discussed three possible actions such as regulatory legislation, business modelling, and agricultural intelligence that can reduce the impact of major disasters or natural disasters in agriculture.

[14], in this paper, authors worked on digital revolution in agriculture. In their work they explained about the important of information technology on agriculture.

[15], in this paper, the authors studied the importance of smart farming and in their work, they cover advanced system models with a large amount of data both in research and business conditions in an effort to bridge the gap between farm systems and large-scale data usage.

In [16], authors studied an advanced agriculture techniques, from the results of their experiments, it is clear that the proposed method produces more visual interpretations of visual effects than other methods in the text. Global food availability and food sustainability is followed by ICT solutions that can be integrated and collaborated as agricultural knowledge. They focused that farmers need to progress themselves to take on smart technologies for farming.

[19], in this paper, author explained various methods for agricultural sustainability, on this work better use of technologies to decrease use of pesticides, environment control by balancing carbon percentages, better irrigation method has been described.

[20], in this paper there is a study done where authors explained that fuel efficient configurations consume less irrigation through the smart use of renewable energy sources (RES).

In paper [22], authors developed a smart irrigation method with low cost the objective of their work to automatically control a water engine and select the direction of water flow in a pipe using a soil moisture sensor. Finally, send the farm field information (water side and directional drives) to the user's mobile message and to the Gmail account.

2. Experimental Works

Challenges in agriculture need to be overcome. In this chapter new and simple designed system models have been presented to help farmers for their irrigation purpose. Simple gear train and epicyclic gear arrangements has been attached with the system separately. So the first system model designed is simple gear train arrangement system and the second system model designed is epicyclic gear arrangement system. Here, the designed systems consists of simple mechanisms and due to its simplicity farmers can operate this system easily for their irrigation purpose.

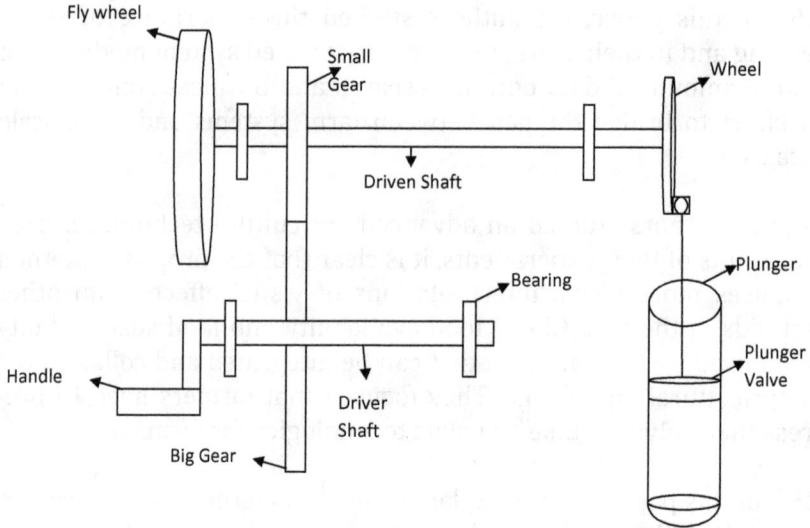

Figure 1. Layout of the Designed Model with simple gear train attachment.

Figure 1 below shows the first model. The model consists of a frame, including some gears, pulley, bearings, shafts, handle. It is required to attach this model with the bore well or tube well. The pulley attached at the end of driven shaft is connected with the plunger of tube well.

Smaller gear attached in driven shaft is meshed with the bigger gear in driving shaft. The driving shaft consists of a handle at its end. The driven shaft consists of a flywheel at its end to store energy and release as required. Number of bearings is there for smooth operation of shafts. Dimensions in this work are taken, may be change as require.

In this work number of teeth on driver and driven gear are taken as 200 and 50. Weight of flywheel is taken as 30 kg. Attachments are connected with a frame, the designed model is very easy to operate, due to its light weight, it can be taken to any places as require. Due to provision of bearings the system operates smoothly.

When effort is applied on handle, it causes to rotate the bigger size gear as attached with the driver shaft. The bigger gear on driver shaft is meshed with the smaller gear on driven shaft. Due to rotation of bigger gear the smaller gear rotate and causes to rotate the driven shaft.

At the end of the driven shaft a wheel is attached, which is connected with the plunger rod of the tube well, due to rotation of wheel the plunger rod moves up and down and results to lift water from tube well. At the other side of the driven shaft a flywheel has been attached which helps for energy storage and dissipate as required.
This is a very easy method and with less effort lifts water from the tube well. So farmers can easily operate this. The designed model is environment friendly due to zero chance of air pollution.

Figure 2. Designed model with frame.

Figure 2 shows the arrangement of the designed model with simple gear train arrangement. The frame used for our work is made up of mild steel angles. So this is the first designed model. In the second designed model, the only difference is of the epicyclic gear arrangement. In the second model, instead of simple gear arrangement, epicyclic gear arrangement is used and both the arrangements are compared later on in the chapter.

Figure 3 shows the arrangement of the epicyclic gear, which consists of sun gear, planetary gear and ring gear.

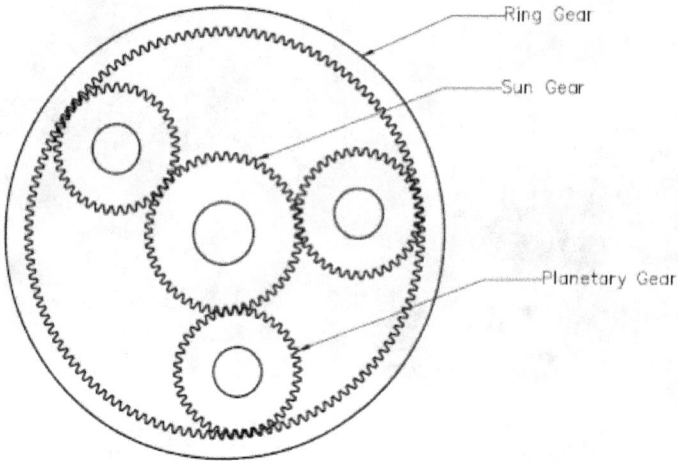

Figure 3. Epicyclic gear train arrangement.

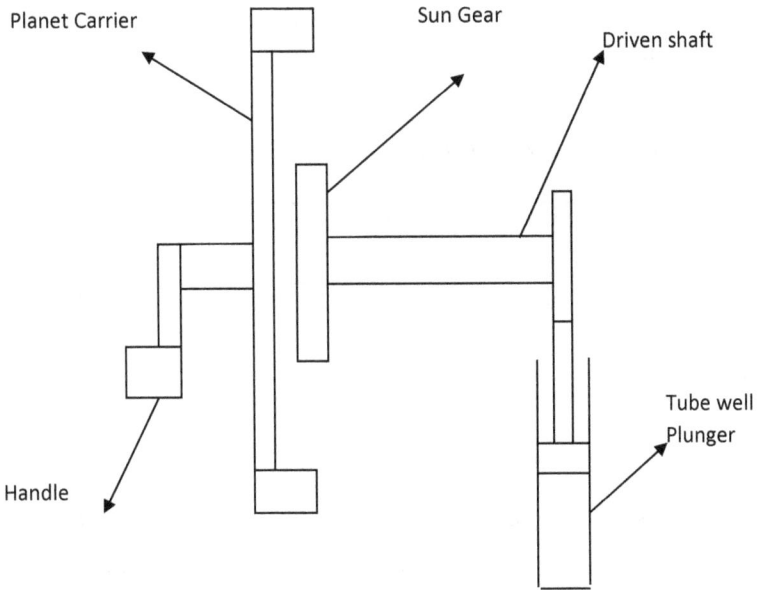

Figure 4. Layout of the designed model with epicyclic gear train attachment.

Figure 4 shows the system with epicyclic gear arrangement. In this arrangement three planet gears are connected with a planet carrier, sun gear is attached with the driven shaft. Driven shaft carries a pulley at its end, the pulley is attached with the plunger rod of the tube well.

Planet gears are attached with a planet carrier which is connected with the handle. When effort applied on handle, planet gears rotate on sun gear, so sun the gear rotates which is connected with the pulley at its end.

The pulley is connected with the plunger rod of the tube well. When sun gear rotates it causes the plunger to have an up and down motion. And finally water lifts from tube well.

As the sun gear is small and power from all three planet gears transferred to sun gear, so in one revolution of handle the sun gear rotate five times, which gives more advantages than first arrangement.

The dimension of the arrangement for epicyclic gear train is given in table 1. In this table dimensions for sun gear, planet gears, ring gears are presented.

Table 1. Specifications of Epicyclic gear arrangement

Sl. No	Specifications	Sun Gear	Planetary Gear	Ring Gear
1	No of teeth	65	55	125
2	Normal module	3.600	3.6000	3.6000
3	Pressure Angle	35 Degrees	35 Degrees	35 Degrees
4	Pitch circle Diameter	145	115	335
5	Addendum	145	120	352
6	Dedendum	6.2925	85.326	322.326
7	Diametral Pitch	25.27	25.27	25.27
8	Circular Pitch	8.231	8.231	8.231
9	Whole Depth	122	7.3925	6.3925

3. Mechanism Behind the Work

A new designed system is used for lifting water from tube well has been presented in this work. Due to progress in technology, there is different mechanism possible which allows to bring advancement in different working areas.

Here, in this work, the designed system consists of gears, shafts, flywheel, bearings and handle. The system is attached with the plunger of tube well. Transfer of motion from handle to plunger causes lifting of water.

Smooth motion of handle is possible due to provision of bearings, flywheel and gear arrangements. Mechanical advantage, velocity ratio, and efficiency concept has been considered in this work.

M.A.= Mechanical Advantage
V.R. = Velocity Ratio
η = Efficiency
W= Load
P= Effort
X= Distance moved by Load
Y= Distance moved by effort

$$M.A. = \frac{W}{P}$$
(1)

$$V.R. = \frac{Y}{X}$$
(2)

$$\eta = \frac{M.A.}{V.R.}$$
(3)

Maximum efficiency is possible for less V.R. and less velocity ratio is possible for maximum value of X (Distance moved by Load) and for minimum value of Y (Distance moved by effort). In this work less effort is provided at handle and maximum plunger motion is happened. So our designed system provides maximum efficiency.

4. Result and Discussion

We have conducted test of our designed system model by taking our designed model in one acre area farm in Cuttack district of Odisha, where vegetable farming was there, there was a tube well with diesel operated pump for irrigation purpose.

We found there was lot of problems with farmers like, transportation of diesel from tank to agriculture field, expenditure for diesel, unavailability of diesel during cyclone periods, etc. Due to such problems their irrigation suffers, and farm output decreases eventually.

To overcome such problems, we install our designed model in that farm with prior permission of farm owner. Both the designed models were tested. After observations, we noticed the difference our both designed model causes and reached a conclusion.

Table 1 below shows the difference noticed and summarises the cost difference between diesel operated pump and our designed models. In our designed system test we have taken factors for comparison as daily expenditure cost, weekly maintenance cost, effect on air pollution, installation, operation factor etc.

Table 2. Comparison between diesel pump and designed models

Sl. No	Factors	Cost with Diesel Pump (2 HP) INR	Cost with Designed Model INR
1	Initial Purchase cost	15,000/-	6000/-
2	Weekly Maintenance	500/-	NIL
3	Effect on air pollution	Major effect	No effect
4	Installation	Skilled Labour	Anyone can install
5	Operation	Chances of risk	Easy to operate

It can be stated from Table 2 that our designed models are easy to operate, easy for installation; there is not any chance of air pollution by using our designed system model. Major advantage of our designed system models is low cost.

It has less initial cost for purchase and maintenance is nil. Due to provision of fly wheel, the designed model works smoothly and farmers can easily use this product for irrigation purpose.

We have also conducted another test of our designed system models in constructional area. This test again confirms that our designed models are very useful. This time our survey was on a 1200 Sq.ft constructional area. We found that from beginning of ground level constructional work to top of the building, water supply was required for betterment of construction.

There were number of labourers engaged for that purpose and they usually lifted water from tube well manually and sprayed on constructional parts. We found that labourers faced a lot of problems for lifting of water. And for same constructional area extra number of labourers were required for lifting of water by the traditional method. Table 3 below shows comparison of requirements using traditional method and with our designed models for 1200 sq.ft constructional area.

Table 3. Comparison of requirements using traditional method and our designed models

Sl. No	Factors	Traditional Method	By our Designed Models
1	No. of Manpower Required	100 people required	40 people required (only for spraying water on building and not for lifting out water from tube well.)
2	Time Requirement	High	Comparatively very less

Table 3 shows the comparison between traditional method of water supply in constructional work with the water supply by developed method. It found that the designed models provides more advantages over the traditional method. This again proved that with our designed system models, lifting water from tube well is very easy. So for water supply, our both designed system models are effective.

Another comparison between simple gear train arrangement and epicyclic gear arrangement is given in Table 4.

Table 4. Comparison between epicyclic and simple gear train arrangement

Sl. No	Factors	Epicyclic Gear train	Simple Gear Train
1	Plunger movement in one complete rotation of Handle	5 times	2 times
2	Weight	Less	More than epicyclic

		arrangement	
3	Effort applied	less	More than epicyclic arrangement
4	Design	Complicated	Simple
5	Maintenance	Difficult	Easy
6	Velocity ratio	less	Slightly more

Table 4 provides some comparison factors between epicyclic and simple gear train arrangements. It was found that the epicyclic gear arrangement provides some more advantages than simple gear train arrangement. In single rotation of handle in epicyclic arrangement 5 times plunger movement is possible, which provides less velocity ratio. According to theory due to less velocity ratio, maximum efficiency is possible for epicyclic arrangement.

5. Conclusion and Scope of Future Work

5.1. Conclusion

In this chapter two designed models for irrigation purpose have been described and tested. Results show that both designed models are very helpful and environment friendly. Our both designed model helps farmers for irrigation purpose and helps labourers in work place for water lifting purpose.

Our designed models are very easy to operate, so farmers can easily operate this for their irrigation purpose. They are environmental friendly, as for its working it doesn't require any external sources.
Our designed models requires less maintenance so farmers will not face problems with our designed models. During its operation, the designed models will not provide any harmful effect, so our product can be use in any time.

As our product will not require any external source (Petrol/diesel/Kerosene/electric), farmers will gain maximum profit from their crop with minimum invest. The second designed model with epicyclic gear train provides some more advantages over simple gear train arrangements. In epicyclic gear train arrangement, it found that in single revolution of handle the plunger rod of tube well provided 5 up and down motion.

5.2. Scope of Future Work

> ➢ The current work is a simple designed model, which has simple gear train and epicyclic gear train attachments with fixed dimensions, but the same may also be conducted by varying dimensions of gears, shafts, plunger length as required.

> ➢ The present work may be tried to be experimented by attaching a wireless sensor network (WSN) consisting of different sensors like soil moisture, temperature and humidit, which can alert farmers about Moisture, Temperature and Humidity condition of farm by sending message.

References

1. Kumar Sahu, C., Behera, P. (2015) A low cost smart irrigation control system, 2nd *International Conference on Electronics and Communication Systems (ICECS)* 2015, pp. 1146-1151.

2. Medici, M., Pedersen, S. M., Canavari, M., Anken, T., Stamatelopoulos, P., Tsiropoulos, Z., Zotos, A., Tohidloo, G. (2021) A web-tool for calculating the economic performance of precision agriculture technology, *Computers and Electronics in Agriculture*, 181, 105930.

3. Martey, E., Etwire, P.M., Mockshell, J. (2021) Climate-smart cowpea adoption and welfare effects of comprehensive agricultural training programs, *Technology in Society*, 64, 101468, .

4. E. M. Lopez, M. Garcia, M. Schuhmacher, J. L. Domingo, (2008) A fuzzy expert system for soil characterization, *Environment International*, 34(7), pp. 950-958.

5. Lioutas, E.D., Charatsari, C. (2021) Enhancing the ability of agriculture to cope with major crises or disasters: What the experience of COVID-19 teaches us, *Agricultural Systems*, 187, pp. 1-5.

6. Nishina, H. (2015) Development of Speaking Plant Approach Technique for Intelligent Greenhouse. *Agriculture and Agricultural Science Procedia*, 3, 9–13.

7. Kawasaki, Y., Matsuo, S., Suzuki, K., Kanayama, Y., Kanahama, K. (2013) Root-Zone Cooling at High Air Temperatures Enhances Physiological Activities and Internal Structures of Roots in Young Tomato Plants. *Journal of the Japanese Society for Horticultural Science,* 82(4), 322-327.

8. Karthikeyan, P., Manikandakumar, M., Sri Subarnaa, D.K., Priyadharshini, P. (2021) Weed identification in agriculture field through iot, *Advances in Intelligent Systems and Computing*, 1163, pp. 495-505.

9. Patil, C., Aghav, S., Sangale, S., Patil, S., Aher, J. (2021) Smart Irrigation Using Decision Tree, *Advances in Intelligent Systems and Computing*, 1245, pp. 737-744.
10. Pavithra, D. S., Srinath, M. S., (2014). GSM based Automatic Irrigation Control System for Efficient Use of Resources and Crop Planning by Using an Android Mobile. *IOSR Journal of Mechanical and Civil Engineering (IOSR-JMCE)*, 11(4), 49-55.
11. Rose, D.C., Wheeler, R., Winter, M., Lobley, M., Chivers, C.-A. (2021) Agriculture 4.0: Making it work for people, production, and the planet. *Land Use Policy*, 100, 104933.
12. Sardini, E., Serpelloni, M. (2011). Self-powered wireless sensor for air temperature and velocity measurements with energy harvesting capability. *IEEE Transactions on Instrumentation and Measurement*, 60(5), 1838-1844.
13. Kansara, K., Zaveri, V., Shah, S., Delwadkar, S. Jani, K. (2015). Sensor based Automated Irrigation System with IOT. *International Journal of Computer Science and Information Technologies (IJCSIT)*, 6 (6), 5331-5333.
14. Satpathy, S. K., Satpathy, S., Chaitanya, P. S., Chowdary, M. M. (2020) Digital Transformation in Agriculture. *Indian Journal of Ecology*, 47, 249-256.
15. Lytos, A., Lagkas, T., Sarigiannidis, P., Zervakis, M., Livanos, G. (2020) Towards smart farming: Systems, frameworks and exploitation of multiple sources, *Computer Networks*, 172, 107147.
16. Tombe, R. (2020) Computer Vision for Smart Farming and Sustainable Agriculture. *2020 IST-Africa Conference (IST-Africa)*, 1-8.
17. Pinto, R., Mathias, C., Kokande, N., Thomas, M., Pushpas, U. S. (2021) Solar Powered Irrigation System, Lecture Notes in Electrical Engineering, 692, 369-381.
18. Goel, S., Sharma, R. (2021) Economic Analysis of Solar Water Pumping System for Irrigation, Lecture Notes in Networks and Systems, 151, 157-167.
19. Pretty J. (2008) Agricultural sustainability: concepts, principles and evidence. *Philosophical transactions of the Royal Society of London. Series B, Biological sciences*, 363(1491), 447–465.
20. Nasiakou, A., Vavalis, M., Zimeris, D. (2016) Smart energy for smart irrigation, *Computers and Electronics in Agriculture*, 129, pp. 74-83.
21. Gutirrez, J., Francisco, J., Villa-Medina, J. F., Nieto-Garibay, A., Porta-Gándara, M. Á. (2014). Automated Irrigation System Using a Wireless Sensor Network and GPRS module, *IEEE Transactions on Instrumentation and Measurement*, 63(1), 166-176.
22. Kumar Sahu, C., Behera, P. (2015) A low cost smart irrigation control system, 2nd International Conference on Electronics and Communication Systems, ICECS 2015, art. no. 7124763, pp. 1146-1151.

23. Pernapati, K. (2018) IoT based low cost smart irrigation system. *Second International Conference on Inventive Communication and Computational Technologies (ICICCT)*, 1312- 1315.

24. Vaishali, S., Suraj, S., Vignesh, G., Dhivya, S., Udhayakumar, S. (2017). Mobile integrated smart irrigation management and monitoring system using IOT. *International Conference on Communication and Signal Processing (ICCSP)*, 2164-2167.

25. Venugoban, K., Ramanan, A. (2014). Image classification of paddy field insect pest using gradient-based features, *International Journal of Machine Learning and Computing (IJMLC)*, 4 (1), 1-5.

Advanced Irrigation Systems in Agriculture: A Literature Review

Manisha Vohra
Independent Researcher, India

Abstract: In agriculture when it comes to growing any type of crops, there are various factors in it which need to be looked after and taken care of like soil condition, climate, compost availability, etc. However, one of the most important factors is irrigation which needs to be carried out on a regular basis, throughout the period when the crop is being grown. To efficiently carry out irrigation in a way that proves to be helpful and eases the work of farmers, various advanced irrigation systems are being talked about, discussed, proposed and even implemented by various experts, professionals, etc. A literature review is carried out in this chapter to understand these advanced irrigation systems in agriculture. A comparison of the reviewed systems is also done in the chapter.

1. Introduction

Agriculture is a very essential sector in our society. It provides us with food which is a necessity, along with many other things like cotton, jute, etc. which are of great use and importance. Human beings cannot live without food. For humans food is a daily requirement thing and is necessary for their survival. Lives of humans greatly rely on agriculture for their food needs.

Food is a daily requirement thing and is necessary for survival of living beings. The source of food is hence of great value and importance. Agriculture being the source of food, the development and progress of agriculture always needs to be looked after and taken care of.

Technology is always looked upon for helping different sectors for their development and progress. Use of technology which could help with different process carried out in agriculture can be helpful. Irrigation is one of the most necessary process in agriculture and technology could help in this process.

2. Irrigation in Agriculture

Agriculture's one of essential part is irrigation. Without irrigation, crops cannot grow and sustain themselves. Things like climatic conditions, compost, etc. do have an impact on crops but irrigation is one of the most important factors for agriculture.

Throughout the duration of crop growth, right from the time the seeds are sowed to the time when crops are fully grown and ready for harvest, this whole duration is time varying which can vary from few weeks to many months, depending on what is being grown. This whole time varying duration, the farmers have to keep vigilance of their crops. They need to monitor them. They have to regularly as per the need and requirement of the crops, carry out the irrigation process in the agricultural field.

Irrigation is necessary for crops survival. However, excess water and less water or no water at all, both are not good for crop's wellbeing. Hence conditions like extremely heavy rainfall, floods and drought are harmful for crop's wellbeing. Irrigation is a compulsory process and part of agriculture. Technology can be put to use for helping with the irrigation process. Thanks to technology, there are advanced irrigation systems being tested for application which we will discussing in literature review.

3. Literature Review

The literature review of different papers is as follows:

3.1 Irrigation System Based on WSN and GPRS

The authors in [10] have developed an irrigation system which is automated. This system is having a wireless network consisting of different sensors, namely the temperature sensors and soil-moisture sensors. These both sensors are deployed in the field where the plants are located. They are deployed in plants root zone [10].

This system is tested in a greenhouse for 136 days. There is a gateway unit that takes care of the information received from, it also triggers the actuators and sends data to web application. For controlling the quantity of water, a microcontroller based gateway was used and it was programmed with an algorithm which was made with threshold values of soil moisture and temperature [10].

Photovoltaic panels were used for powering the system. The inspection of the data and scheduling of the irrigation could be done through the web page. GPRS module was used to transmit the data to web server via the public mobile network [10]. As mentioned earlier, this system has been tested in a greenhouse for 136 days which witnessed the saving of water up to 90% in comparison with traditional irrigation practice [10].

3.2 Irrigation System Based on Machine Learning and WSN

The authors in [14] have developed and implemented a solution for irrigation. They have developed an irrigation control system which is an automatic system. Machine learning plays an important role in this system. The system comprises of wireless sensors and actuators network. The data which is gathered with the help of wireless sensors network is sent using narrowband internet of things to a cloud based server for storing and analysis purpose [14].

The developed system consists of a mobile application which offers users with the ability of not just viewing the data gathered in actual real-time but it also lets them view the history and act according to the data it analyses. The authors try to save water through their solution which is their important aim for developing this solution [14].

Different machine learning algorithms like the Random Forest algorithm, Decision Tress algorithm, etc. were studied for estimating the best time of the day for administrating the water. Random Forest algorithm gave the best results. It gave an accuracy of 84.6%. Along with the machine learning algorithm based solution, a method was made to find out how much water is required for managing the fields which was under analysis. This system on implementation can allow for savings of water till 60% [14].

3.3 Irrigation System Based on ANN Based Controller and Evapotranspiration Model

In this paper [15], the authors have proposed artificial neural network (ANN) based controller for irrigation. In this system different

sensors are used to get soil moisture, temperature, humidity and radiation. Also used is a Global Positioning System (GPS) in this proposed system. MATLAB is used for this system prototyping [15].
The desired soil moisture value is obtained from database of the agricultural research center. Using the actual values of different parameters like soil moisture, temperature, etc. received from the sensors, required soil moisture can be calculated from the Eto by comparing desired and actual soil moisture values. Eto is reference evapotranspiration [15].

Using ANN, the authors are getting accurate results when calculating required soil moisture from the input parameters. The irrigation process decision on the basis of sensor values and its comparison with the standard values can be decided automatically and the authors have the option of even deciding manually by passing commands for switching on and off for the motor [15].

3.4 Irrigation System Based on Soil Moisture Sensor and Raspberry-Pi

The authors in this paper [16] have proposed and put forth an automated irrigation control system prototype. Soil moisture sensor is used. In the system prototype, there is sensor node and control node present, out of which the sensor node is present in the field. It senses the soil moisture. The data is then sent to controller node which will see the received value of the moisture and check it [16].

The received value of soil moisture will be checked and compared against the needed value. If the received value of the soil moisture is less that the needed value then it means irrigation needs to be started automatically by turning on the motor. An electromagnetic valve is used which is controlled by Raspberry-Pi. In whichever direction the field is dry or having less moisture as compared against the needed value, the water is provided there, thus automatically starting irrigation and taking care of irrigation [16].

Also, when irrigation starts, an alert message is sent to the mobile number which is registered in the system and also the message is sent to Gmail account. This system prototype was tested and it was found that it works as desired. It is able to automate the irrigation and it is helpful in saving energy [16].

3.5 Irrigation System Based on WSN and GPRS Module

In this paper [17], the authors have proposed a smart irrigation system. It uses wireless sensor network along with General Packet Radio Service (GPRS) network for the proposed system. Here the irrigation system is proposed for the case of rice crops [17].

There are different sensors like temperature sensor, water level sensor, humidity sensor and water level sensor used in this system. Also, GPRS, microcontrollers and zigbee modem is used in the system [17]. When the threshold values of temperature and soil moisture sensors are reached, irrigation is started automatically. The automated irrigation system implemented was found to be feasible and cost effective for optimizing water resources for agricultural production [17].

4. Comparison of Literature

Paper No. in references	Year	Authors	Irrigation system based on	Result obtained	Advantages
[10]	2014	Gutirrez, J., Francisco, J., Villa-Medina, J. F., Nieto-Garibay, A., Porta-Gándara, M, Á.	WSN and GPRS module	This system has been tested in a greenhouse for 136 days which witnessed the saving of water up to 90% in comparison with traditional irrigation practice.	Irrigation could be automated and up to 90% water savings was there as compared to traditional practice of irrigation.
[14]	2021	Glória, A., Cardoso, J., Sebastião, P.	Machine learning and wire-	Random Forest algorithm	A system for automatic irri-

			less sensor network	for estimating the best time of the day for administrating the water gave an accuracy of 84.6% and WSN for gathering and viewing data gathered in actual real-time. Finally with a method for finding out required water amount, 60% water savings were achieved.	gation control is developed and great amount of water savings is achieved.
[15]	2013	Raja Sekhar Reddy. G, Manujunatha. S, Sundeep Kumar K.	ANN based controller and evapotranspiration model	The irrigation process decision on the basis of sensor values and its comparison with the standard values can be decided automatically or even manually by	Using ANN, accurate results are obtained when calculating required soil moisture from the input parameters.

				passing commands for switch- ing on and off for the motor.	
[16]	2015	Sahu, C. K., Behera, P.	Soil mois- ture sen- sor and Rasp- berry-Pi	When soil moisture is less as compared against the needed value, irri- gation starts au- tomati- cally and an alert message is sent to mo- bile and Gmail ac- count.	It works as desired and it is helpful in saving en- ergy.
[17]	2015	Sreeni- vasulu, P., Harinath, V., Ramaiah	Wireless sensor network and GPRS module	Irrigation starts au- tomati- cally when threshold values of tempera- ture and soil mois- ture sen- sors are reached.	Feasible and cost effective system.

5. Discussion

A literature review was carried out in which we discussed some advanced irrigation systems. A comparison of them was also conducted.
It is seen that various different advanced irrigation systems are coming across as highly useful. Their results are encouraging. It can be

clearly seen and understood from this literature review and the results of the systems stated in them, that here in the literature review, the irrigation systems which are proposed by the different authors, can be of great help and advantageous in agriculture for carrying out irrigation process.

6. Conclusion

A literature review of different advanced irrigation systems in agriculture has been presented in this chapter in order to understand the different advanced irrigation systems in agriculture and a comparison of the reviewed systems has been carried out and presented as well in this chapter.

Agriculture has different requirements like it requires compost, irrigation, etc. In fact, in agriculture, irrigation is not just a requirement but also one of the very vital factors. Irrigation is must in agriculture. On the basis of the results for the proposed or suggested advanced irrigation systems obtained by the authors of the papers reviewed in this chapter, it can be stated that the advanced irrigation systems can be helpful in easing out work of the farmers by enabling automation of irrigation and can even provide advantages like saving of water, etc. Altogether, it can be clearly understood from this review that advanced irrigation systems can be useful.

References

1. Venugoban, K., Ramanan, A. (2014). Image classification of paddy field insect pest using gradient-based features, *International Journal of Machine Learning and Computing (IJMLC)*, 4 (1), 1-5.
2. Mekala, M. S.; Viswanathan, P. (2017) A novel technology for smart agriculture based on iot with cloud computing. *In Proceedings of the International Conference on IoT in Social, Mobile, Analytics and Cloud (I-SMAC)*, pp. 75–82.
3. Kawasaki, Y., Matsuo, S., Suzuki, K., Kanayama, Y., Kanahama, K. (2013). Root-Zone Cooling at High Air Temperatures Enhances Physiological Activities and Internal Structures of Roots in Young Tomato Plants. *Journal of the Japanese Society for Horticultural Science*, 82(4), 322-327.
4. Kansara, K., Zaveri, V., Shah, S., Delwadkar, S. Jani, K. (2015). Sensor based Automated Irrigation System with IOT. *International Journal of Computer Science and Information Technologies (IJCSIT)*, 6 (6), 5331-5333.

5. Nishina, H. (2015). Development of Speaking Plant Approach Technique for Intelligent Greenhouse. *Agriculture and Agricultural Science Procedia*, 3, 9–13.

6. Liakos, K. G., Busato, P., Moshou, D., Pearson, S., Bochtis, D. (2018). Machine Learning in Agriculture: A Review. *Sensors*, 18(8), 2674, 1-29.

7. Sardini, E., Serpelloni, M. (2011). Self-powered wireless sensor for air temperature and velocity measurements with energy harvesting capability. *IEEE Transactions on Instrumentation and Measurement*, 60(5), 1838-1844.

8. Davis, S. L., Dukes, M. D. (2010). Irrigation scheduling performance by evapotranspiration-based controllers. *Agricultural Water Management*, 98 (1), 19-28.

9. Pavithra, D. S., Srinath, M. S., (2014). GSM based Automatic Irrigation Control System for Efficient Use of Resources and Crop Planning by Using an Android Mobile. *IOSR Journal of Mechanical and Civil Engineering (IOSR-JMCE)*, 11(4), 49-55.

10. Gutirrez, J., Francisco, J., Villa-Medina, J. F., Nieto-Garibay, A., Porta-Gándara, M. Á. (2014). Automated Irrigation System Using a Wireless Sensor Network and GPRS module, *IEEE Transactions on Instrumentation and Measurement*, 63(1), 166-176.

11. Rajalakshmi, P., Mahalakshmi, S. D., (2016). IOT based crop-field monitoring and irrigation automation, *2016 10th International Conference on Intelligent Systems and Control (ISCO)*, 1-6.

12. Vaishali, S., Suraj, S., Vignesh, G., Dhivya, S., Udhayakumar, S. (2017). Mobile integrated smart irrigation management and monitoring system using IOT. *International Conference on Communication and Signal Processing (ICCSP)*, 2164-2167.

13. Pernapati, K. (2018). IoT based low cost smart irrigation system. *Second International Conference on Inventive Communication and Computational Technologies (ICICCT)*, 1312- 1315.

14. Glória, A., Cardoso, J., Sebastião, P. (2021). Sustainable Irrigation System for Farming Supported by Machine Learning and Real-Time Sensor Data. *Sensors*, 21, 3079, 1-26.

15. Raja Sekhar Reddy. G, Manujunatha. S, Sundeep Kumar K. (2013). Evapotranpiration Model Using AI Controller for automatic Irrigation system. *International Journal of Computer Trends and Technology (IJCTT)*, 4(7), 2311-2315.

16. Sahu, C. K., Behera, P. (2015). *A Low Cost Smart Irrigation Control System. 2015 2nd International Conference on Electronics and Communication System (ICECS 2015)*, 1146-1152.

17. Sreenivasulu, P., Harinath, V., Ramaiah, G. N. K. (2015). Smart Irrigation System using a Wireless Sensor Network and GPRS Module. *International Journal of Scientific Engineering and Technology Research (IJSETR)*, 4(6), 1019-1023.

18. M. Pagliai, N. Vignozzi, S. Pellegrini, (2004) Soil structure and the effect of management practices, *Soil and Tillage Research*, Vol. 79, No. 2, pp. 131-143.

19. Patil, C., Aghav, S., Sangale, S., Patil, S., Aher, J. (2021) Smart Irrigation Using Decision Tree, *Advances in Intelligent Systems and Computing*, 1245, pp. 737-744.

20. Nasiakou, A., Vavalis, M., Zimeris, D. (2016) Smart energy for smart irrigation, *Computers and Electronics in Agriculture*, 129, pp. 74-83.

21. López, E. M., García, M., Schuhmacher, M., Domingo, J. L. (2008) A fuzzy expert system for soil characterization, *Environment International*, 34(7), pp. 950-958.

Crop Yield Prediction for Improving Production in Agriculture

Harismithaa, L. R. and Prakash, J.
Department of CSE, PSG College of Technology, Coimbatore, India

Abstract: Agriculture forms a significant part of every country's economic growth. It has been a field of deep research in the past and it continues to be. Many researchers are carrying out research in the field of agriculture in improving the betterment of agricultural production. To predict the yield of agricultural crops machine learning models can be used in prediction based on parameters like weather data, nature of the soil at the crop cultivation environment. In this chapter, the prediction models like random forest and decision trees are used to predict the yield of crops in advance based on the weather, soil and historic crop yield data, which will help the farmers in identifying the suitable crop for cultivation. The performance of the prediction models was evaluatedon the crop yield data of maize collected for different seasons and the results inferred that random forest model outperforms decision tree models in predicting the crop yield.

1. Introduction

Agriculture plays a predominant role in a country's economic progress. With the continuing expansion of the human population, understanding worldwide crop yield is vital to address the food stock & supply challenges and reducing the impacts of climate change. The crop yield prediction has gained significance worldwide and is one of the most challenging tasks for every country.

Agricultural efficiency and production are fundamentally determined by the inputs, investment and methods of production employed. Progressive agriculture demands improvements in cropping methods and smart cultivation patterns.

Technology stands first among all the factors determining the production in agriculture.The role of technology in agricultural improvement

is to make proper use of agricultural inputsfor devising new impro-
vised cultivation patterns in order to increase the production. It aims
to make rational use of the scarce land-resource by effectively predicting in advance, the yield ofthe crop that can be obtained. Besides
this, weather conditions are also a concern in crop production. Based
on the weather conditions such as temperature, Rainfall the yield may
vary.Adverse climatic conditions affect the production to a greater
extent. With the help of crop yield prediction in advance the cultivation can be done in most economical manner without any waste of
man power and capital investment [12].

In the recent years, data analytics has boomed into action in every
field improving the standards of living by providing prediction techniques and risk assessment [13-16]. It helps by analyzing the factors
influencing the yield and making predictions on the harvest. Machine
Learning is a part of data analytics that renders various algorithms for
classification, regression, and risk assessment. Estimating the production that can be expected from the cultivation will benefit the
farmers in making appropriate decisions. The advance prediction of
the yield also benefits the policy makers for fixing the market price of
the crops and to take suitable processesfor storage of the crops and
shipping to the end customers [17].

The objective of the chapter is to predict the yield of crops using different machine learning algorithms. Decision trees and Random Forests algorithms are used for predicting theyield and their performance is compared. The prediction results significantly support the
farmers in taking up appropriate decisions in cultivation thus improving the production in the agricultural sector. Section II describes
about the existing works by the researchers on crop yield prediction.

The remainder of the chapter is structured as, related study in section
2 and section 3 describes the data used. Section 4 discusses about the
prediction models for crop yield prediction, section 5 presents the
performance metrics suitable for evaluating the models for the crop
yield prediction. Section 6 compares and analyses the results obtained
by the classifiers followed by conclusion in section 7.

2. Related Work

P. Priya et al., [1] (2018) proposed an approach for predicting production of agricultural crops usingmachine learning algorithms. The study used the crop data collected from different districts of Tamil Nadu. The crop yield was predicted from the past data of weather, soil type and crop yield of Tamil Nadu. This paper implemented Random Forest algorithm in R for prediction ofthe yield of the crop and an accuracy of about 56% was obtained.

A. Nigam et al., [2] (2019) proposed a methodology for crop yield prediction using machine learning algorithms which included data of temperature, rainfall and production of crops collected from various sources and then combined into a final dataset. The paper used factors such as temperature and rainfall for the crop yield prediction. The study used differentensemble classifiers such as Random forests, XGBoost, K - Nearest Neighbour classifier and Linear Regression. Results reveals that Random Forest is the best classifier when all parameters such as temperature, rainfall, season, area and production details are combined.

S.S. Kale et al., [3] (2019) proposed a study on various machine learning approaches to predict the yield of crops and success rate of the prediction in reality. The dataset includes experimental parameters such as are cultivation area, crop, state, district, season, year and production or yield for the period of 1998 to 2014 for Maharashtra state. This research developed a 3 Layer Perceptron ANN model with ReLU activation function for different crop yield prediction. The paper used production as output parameter and features like crop, area, district, season as the input to the Multi-layer Perceptron. The Multilayer perceptron model was implemented in Python Tensor Flow. The model obtained 80% accuracy for the input dataset.

M. Champaneri et al., [4] (2020) presented a study on crop yield prediction using machine learning by building a prototype of an interactive prediction system with a web basedgraphic user interface. The data for the study was collected from each district of Maharashtra. The dataset comprised of climatic measures like precipitation, temperature, cloud cover, vapour pressure, wet day frequency and district level crop data of the districts of Maharashtra.The dataset was partitioned as 75%:25% for the training and testing phase. The paper used a random forest classifier for prediction and an accuracy of about 75% was obtained for all the districts and crops selected in the study.

Sangeeta et al., [5] (2020) performed a comparative study on design andimplementation of different machinelearning classifier models in agricultural yield. The proposed approach analyses several atmospheric aspects like temperature, rainfall, soil pH, humidity, type of soil and production of the crop in previous year's crop production and predicts the best produced crop for a particular region. The paper uses Random forest, Polynomial Regression and Decision Tree models. The paper compares the performance of the three machine learning models and evaluated based on accuracy and precision metrics. Random forest classifier obtained highest accuracy of 88% among the classifiers.

Y.J.N. Kumar et al., [6] (2020) presented a study on supervised machine learning approach for crop yield prediction. The study involved factors like rainfall, humidity, temperature, crop name and production in previous years. The paper appliedDecision tree algorithm and Random forest classifier and their results were compared based onthe accuracy and precision metrics. The findings of the paper are Random forest classifier obtained higher accuracy and also reduced the over fitting problem.

Table 1 summarizes the study of various approaches carried out in field of crop yield prediction using machine learning models.

Table 1. Summary of Related works on crop yield prediction

Author	Year	Learning Models	Results and Findings	Data Used	Scope for Improvement
P. Priya et al.,	2018	Random Forest	Accuracy obtained in this approach is 56%	Crop data collected from different districts of Tamil Nadu, India.	The accuracy can be further improved
A. Nigam et al.,	2019	Ensemble classifiers such as Random forests, XGBoost, KNN classifiers, Logistic Regression and Linear Regression.	Random forests provided better accuracy of 67% when compared to other models	Temperature, Rainfall and Crop production data across all the districts in India	Highest Accuracy obtained by Random forest classifier model in the paper is 67%. Thus, the accuracy can be further improved.
S.S. Kale et al.,	2019	ANN-Multi Layer Perceptron with 3 layers	The model obtained 80% accuracy.	Crop data collected for the period of 1998 to 2014 for the state of Maharashtra, India.	The ANN algorithm can be improved further to give more accurate prediction by adding more layers.

M. Champaneri etal.,	2020	Random forestclassifier	Accuracy of about 75% wasobtained for allthe districts and crops selected	District level crop data of the state of Maharashtra, India.	The accuracy can be improved further and can be implemented for prediction in other states.
Sangeeta et al.,	2020	Random forest,Polynomial Regression, Decision Tree.	Random forestclassifier provides betteryield prediction with 88% testing accuracy	State wise agricultural data in India	Future improvements canbe integrating the model with other departments like horticulture and sericulture.
Y.J.N. Kumar et al.,	2020	Decision treesand Random forestclassifier	Random forest reduced over fitting problemand obtained accuracy of 86%	Temperature, Rainfall, Humidity, ph data collected from different States of India.	This system worksfor structured data. It can be extended for unstructured data.

3. Data Used for Prediction Models

Crop Yield Prediction is based on the available historical data like weather parameters, soil parameters and historic crop yield of maize, collected during seasons- summer, autumn and winter. Figure 1 shows the sample view of data used for the prediction.

	Season	Temperature	Rainfall	pH	Area	Production
0	2	29.01000	251.54975	5.6	4.0	7.5
1	2	28.37825	248.81725	6.6	3.0	5.6
2	2	29.33725	144.16450	5.4	3.0	14.0
3	2	30.50325	226.06250	6.2	4.0	6.0
4	2	20.44425	424.34625	5.1	257.0	676.0

Figure 1. Sample view of the crop yield data

The different parameters used are:
- Season
- Rainfall
- Temperature
- Area of cultivation
- pH of the soil
- The corresponding maize production for the parameters.

These parameters in the data are described in Table 2.

Table 2. Description of the data attributes in the crop yield data

Sno.	Parameter	Description	Range of values
1	Season	It refers to the seasons of the year spanning from 3-4 months in a year	Summer, Autumn and Winter
2	Temperature	It refers to the temperature recorded over a period of time in Celsius	18-35 Celsius
3	Rainfall	It refers to the amount of rainfall recoded in the particular season measured in mm	0.6-389 mm
4	pH of the soil	It refers to the nature of the soil which can be acidic, alkaline or neutral on the pH scale (0-6 refers to acidic nature , 8-14 refers to alkaline nature and 7 pertains to neutral nature of the soil)	0 - 14
5	Area of Cultivation	It refers to the area of land subjected to the cultivation of maize of a period of time measured in hectares	1-200 hectares
6	Production of Maize	This parameter specifies the production obtained for the given parameters measured in Thousands of INR	INR 1000 – 1,37,680

4. Prediction Models

The proposed method implements two classification algorithms Decision trees and randomforests for predictions on the crop yield data and their performance is analyzed in terms of the performance metrics selected.

Decision Trees

Decision trees is a supervised learning model widely used for predictive modelling in statistics, data mining and machine learning. It can be used for classification and regression. Decision trees typically take the structure of the tree. The internal nodes of the decision trees pertain to the input features of the decision problem. The internal nodes are connected hierarchically through branches which correspond to the decision rules. The decision rules endup in leaf nodes which represent the decision or target label based on the decision rules(branches) learnt from input features (internal nodes).

Decision tree algorithm is implemented for the crop yield data using sklearn package. The train-test split of the data is in a ratio of 7:3. The train_test_split() function in sklearn is used to split the data in the specified ratio. The Decision Tree Regressor() with maximum tree depth as 25 and maximum features in each level as 5 is trained with the training data. The model is tested for predictionson the test data. The Decision tree regressor obtained an accuracy of 87.5%.

Random Forest

Random forest is also a powerful supervised learning model which performs both classification and regression tasks. It employs a concept of ensemble learning where constructsa multiple decision trees to provide better prediction.

The number of decision trees are built upon the subset of the input data and finally the decision from each decision tree are averagedto improve the prediction accuracy. Based on the majority of the votes of the prediction, the final decision or the target variable is predicted. The greater the number of trees in Random forest, the prediction accuracy increases and also prevents the problem of over fitting.

Random Forest algorithm for the crop yield data is executed using the functions providedby the sklearn package. The dataset is split in a ratio of 7:3 for training and testing using train_test_split() function in sklearn. The Random Forest Regressor() with 100 estimators is trained with the training data. The model is tested for predictions of the yield. The Random forest regressor obtained an accuracy of 90.1% for the yield predictions.

5. Performance Metrics

Accuracy

Accuracy is used as the performance metrics for the crop yield prediction system sincetrue positives and true negatives are more significant in the prediction of yield. Accuracy is thenumber of data entries predicted correctly among all the data entries. The number of data entriespredicted correctly corresponds to the sum of all the true positives and the true negatives as shown in Equation (1).

$$Accuracy = \frac{True\ Postives + True\ Negtives}{True\ Positives + True\ Negatives + False\ Positives + False\ Negatives} \quad (1)$$

6. Result Analysis

The Random forest model exhibits better training and testing accuracy compared to Decision trees. Random Forest takes a number of decision trees on numerous subsets of the given data and selects the best decision from using ensemble voting scheme. From Table 3,it can be deduced that Random forests outperforms decision trees by difference of 2.6% in accuracy metrics. Thus it can be recommended for crop yield prediction system.

Table 3. Accuracy obtained by the classifiers

Models	Accuracy
Decision Tree	87.5%
Random Forest	90.1%

The following figure (Figure 2) shows the accuracy of the random forest and decision trees in prediction of the maize yield. The training accuracy of 87.5 and 90.1 percent were obtainedby the decision tree and random forest classifiers respectively with number of estimators as 20.

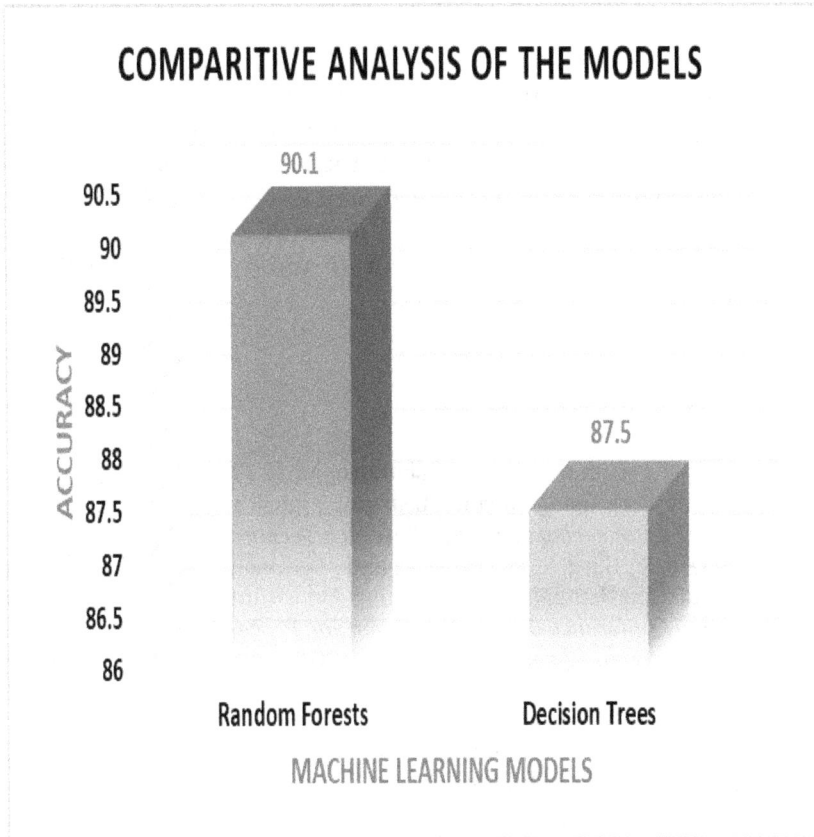

Figure 2. Comparison of machine learning models based on the accuracy metrics

7. Conclusion

This chapter implements crop yield prediction system for predicting the yield of the required crop given the necessary input climatic, area and soil parameters. The study aims to make rational use of the scarce land resources by analyzing the various factors associated with the cultivation.

The predictions are carried out on the input data using machine learning algorithms such as Random Forest and Decision Tree algorithm. The results show that Random Forest exhibits better training and testing accuracy when compared to Decision Trees for the given input data.

The prediction intends to help the farmers and the retailers to predict the production and cost of the agricultural commodities in advance. This work can be extended to predict the yield of agricultural crops by adding specific parameters like ground water level, previous crop patterns cultivated in that area, Soil moisture content, soil fertility level to the input data for more accurate and detailed prediction of the yield.

References

1. Priya, P., Muthaiah, U., and Balamurugan, M. (2018) Predicting Yield Of The Crop Using Machine Learning Algorithm. *International Journal of Engineering Sciences &Research Technology (IJESRT).*
2. Nigam, A., Garg, S., Agrawal, A., and Agrawal, P. (2019) Crop Yield Prediction UsingMachine Learning Algorithms. *IEEE Fifth International Conference on Image Information Processing (ICIIP).*
3. Kale, Shivani S., and Patil, Preeti S. (2019) A Machine Learning Approach to Predict Crop Yield and Success Rate. *IEEE Pune Section International Conference (PuneCon).*
4. Champaneri, M., Chandvidkar, C., Chachpara, D., and Rathod, M. (2020) Crop Yield Prediction Using Machine Learning. *International Journal of Science and Research.*
5. Sangeeta., and Shruthi, G. (2020) Design And Implementation Of Crop Yield Prediction Model In Agriculture. *International Journal of Scientific & Technology Research(IJSTR), vol. 8, issue 01.*
6. Jeevan Nagendra Kumar, Y., Spandana, V., Vaishnavi, V. S., Neha, K., and Devi, V.
7. G. R. R. (2020) Supervised Machine learning Approach for Crop Yield Prediction in Agriculture Sector. *IEEE Fifth International Conference on Communication and Electronics Systems (ICCES 2020).*
8. Medar, R., Rajpurohit, Vijay S., and Shweta. (2019) Crop Yield Prediction using Machine Learning Techniques. *IEEE 5th International Conference for Convergence inTechnology (I2CT).*
9. Guruprasad, R. B., Saurav, K., Randhawa, S. (2019) Machine Learning Methodologiesfor Paddy Yield Estimation in India: A Case Study. *IEEE IGARSS.*
10. Veenadhari, S., Misra, B., and Singh, C. D. (2014) Machine learning approach for forecasting crop yield based on climatic parameters. *IEEE International Conference onComputer Communication and Informatics (ICCCI).*
11. Sharma, B., Yadav, S., and Pratap Singh Yadav, J. K. (2020) Predict

Crop Production in India Using Machine Learning Technique: A Survey. *IEEE 8th International Conference on Reliability, Infocom Technologies and Optimization (ICRITO).*

12. Manjula Josephine, B., Ruth Ramya, K., Rama Rao, K.V.S.N., Kuchibhotla, S., Venkata Bala Kishore, P., and Rahamathulla, S. (2020) Crop Yield Prediction Using Machine Learning. *International journal of scientific & technology research, vol 9, issue 02.*

13. Sivanandhini, P., & Prakash, J. (2020). Comparative Analysis of Machine Learning Techniques for Crop Yield Prediction. *International Journal of Advanced Research inComputer and Communication Engineering, 9(6), 289-293.*

14. https://doi.org/10.17148/IJARCCE.2020.964

15. R Sandhya, J Prakash, B Vinoth Kumar (2020). Comparative Analysis of Clustering Techniques in Anomaly Detection in Wind Turbine Data. *Journal of Xi'an Universityof Architecture & Technology, Vol. 12 No.3 2020, pp. 5684-5694.*

16. Prakash, J., & Joy, E. J. (2020). A comparison of different surrogate models for delamination detection in composite laminates using experimental modal analysis. *Proceedings of Advanced Materials, Engineering & Technology.* https://doi.org/10.1063 /5.0019366

17. Prakash J. (2018). Enhanced Mass Vehicle Surveillance System, *J4R, Volume 04, Issue04 ,002, 5-9, June2018.*

18. Rajesh, R., &Mathivanan, B. (2017). Predicting Flight Delay using ANN with Multi-core Map Reduce Framework. *In Communication and power engineering (p. 280). Walter de Gruyter GmbH & Co KG.*

19. Sivanandhini, P., & Prakash, J. (2020). Crop Yield Prediction Analysis using Feed Forward and Recurrent Neural Network. International Journal of Innovative Science and Research Technology, 5(5), 1092–1096. doi:10.38124/volume5issue5.

20. Harsimithaa, LR and Sudha Sadasivam, G. (2021). Multimodal screening of Dyslexia using Handwriting and Eye gaze. *Journal of Huazhong University of Science and Technology (JHUST), vol.50 no.4.*

Agricultural Plant Leaf Disease Identification and Classification Using Deep Learning Approach

[1]Anilkumar Suthar, [2]Jashraj Karnik and [3]Ramesh Prajapati
[1]New LJIET, Ahmedabad, Gujrat, India
[2]Department of Computer Engineering, LJIET, Ahmedabad, Gujarat, India
[3]Department of Computer Engineering, SSIT, Ahmedabad, Gujarat, India

Abstract: Plant leaf disease is a major problem in the agricultural sector. In various seasons, several plant leaf diseases appear. A deep learning approach, which can help farmers understand the plant leaf disease and identify them can be used. In the area of research that involves the feature extraction for plant leaf disease using image processing techniques can help farmers make optimal decisions quickly and accurately. A study has been carried out in this chapter and on its basis the authors have proposed a system where they have used image pre-processing and image augmentation techniques to obtain a better image, which allow them to process better results for analysis. In this study, three types of leaves take into consideration, namely bell pepper, potato and tomato leaves. Classification is performed in the proposed system for classifying plants leaf disease of bell pepper, potato, and tomato leaves. The system is divided into two stages. In the first stage, classifier, a median filter and image augmentation are used and the algorithm is trained and second stage classifier, the extract plant leaf image output is performed using ResNet50. So, with two step classification approaches and a deep learning approach to respective plant leaves for observing the diseases, 98% accuracy for detecting and classifying the leaf disease.

1. Introduction

Agriculture is a very important profession. In the economic sector also, it is very important. Worldwide, agriculture is a huge industry. The global population is expected to reach rise, which will require an increase in agricultural production to fulfil the demand. The production of crops in agriculture can be increased and well managed by using efficient methods. A crop in simple words is a plant or its product. It can be grown and harvested, as well. Various types of crops are grown in the world. There is a lot of variation in environmental conditions in different regions of the world. Temperature, humidity, rainfall, all of them vary greatly in different places. Pesticides if used excessively then can have ill effects on the plant and also it is very important to see that the plant is not affected by any disease. So the right amount of pesticides should be used. If in case disease occurs, then it is essential to straightaway detect the disease. The plant diseases detection and identification is required to maximize production rate and thus the income being generated from it. The best solution to the problem is to identify the disease of the plant so that precautionary steps can be taken to safeguard the same.

Many researchers have used data mining techniques or machine learning technique or deep learning techniques to predict the possible outcome. The important part in the agricultural field is yields, yields are nothing but the area of a particular sector or land fixed to measure the outcome. The yields only increase when number of defects are less in the fields i.e. by checking the plant leaf, every time. This machine learning or deep learning is a basic process using the various algorithms for prediction of soils, weathers, plant leaves, growth of plant, identify the seed quality etc. in agriculture sectors. In the field of agriculture, plant leaf detection and classification are used to classify in which image is to classify the true label, the problem with classification is simply classifying too many objects simultaneously or a single object. The convolutional neural network has shown vast improvement in the algorithm in classification in these domains. So, in order to classify more deeply the detection algorithm is used to detect the multiple regions of picking more than one object in a single image.

Image classification is core problem in computer vision with practical applications. Image classification is the task of assigning an input image, one label from the fixed set of categories. Image classification which is used to classify the image into classes by labels. It takes the input and outputs the classification label of that image with some metrics (loss, accuracy, bias, pooling, etc.). Convolutional Neural Network (CNN) has been state of art algorithm which mostly used for classification approach and has more neuron (i.e., hidden layer) than

ANN [22]. So, Convolutional Neural Network (CNN) has many parameters and has different feature to improve classification process.

1.1 Need of Deep Learning in Agricultural Sector

Agriculture is one of the important sectors in the society. Different technological innovations have brought different options and opportunities for farmers. Many new technologies, such as Machine Learning and Deep Learning, are being implemented in agriculture so that it can benefit the agricultural sector.

Agriculture is an area where there can be a huge impact of Artificial Intelligence (AI) based deep learning or machine learning methods. Agriculture is impacted by a multitude of variables which includes crop specifications, soil conditions, climate change, etc. Deep learning can help in agriculture in various ways like in plant leaf disease identification and classification, monitoring crop yield, etc.

1.2 Application of Deep Learning in Agricultural Area

- Analyzing crop health
- Identifying plant leaf disease
- Classifying plant leaf disease

2. Plant leaf Disease Detection

In farming process, the most widely used practice in pest control and disease monitoring for protection of crops is regular spray of pesticides. For this approach to be effective, it requires significant amounts of pesticides which results in an increase in overall financial cost moreover, it can affect the crop health if sprayed in excess quantity.

Below is the list of different general challenges which already exist in the agricultural sector.

➢ In farming climatic factors such as rainfall, temperature and humidity play an important role in the agriculture lifecycle. Increasing deforestation and pollution result in climatic changes can affect the soil and harvest.

➢ Every crop requires specific nutrition in the soil. There are 3 main nutrients Nitrogen (N), Phosphorous (P) and Potassium (K) required for soil to maintain the fertility. Nutrient deficiency can result in having poor quality crops.
➢ Another challenge in crops we can see from the agriculture lifecycle that seed protection plays an important role. If it is not controlled, it can increase the cost of production. It can also absorb out different nutrients from soil that can lead to it can lead to deficiency of nutrition in soil.

As a solution to the said problem, below are the summarized uses of deep learning for analysis, and to detect and classify any disease that might have affected a plant by taking an image of the leaf. The processing pipeline goes as follows:

1. The leaf is detected in the given image and cropped out at each stage of growing plants.
2. The extracted leaf is then run through a classifier to identify which plant the leaf belongs.
3. The leaf is then checked for the disease class, if any, based on the result of the previous step.

Deep learning (DL) is a machine learning subset that enables the learning of hierarchical data. The fundamental benefit of DL is that features automatically learn from input data, which removes hurdles for the creation of smart solutions to various applications. Deep neural networks (CNNs) are frequently employed DL architectures. The latest in performance has been reached with major vision applications including image classification/regression, object identification, and picture segmentation (both semantic and instances) [23, 9]. CNNs began in the 1980s [1] and demonstrated their initial success with back-propagation in the recognition of hand-written numbers in the 90's.

2.1 CNN Based Image Processing Methods

CNN's undergo different changes very fast. That is why, this study is only confined to models with considerable increase in performance and frequently utilized in other domain applications as benchmarks [22].

An overall breakthrough (Alexnet) was realized in 2012 since the computational strength (and consequently complexity of the CNN model) was significantly improved and the images (e.g. ImageNet) were available [2]. CNN's for other vision tasks such as object identification, semantic and instance segmentation, have also been frequently employed as feature extractors with meta-architecture. In compared to conventional techniques for practically all of these tasks, the CNNs produced state-of-the-art results, which demonstrate a strong potential to increase the performance of data processing in such imaging-based applications as imaging-based plant physical composition. The advances of the transfer learning (a technique that helps to transfer features learned from one data set to the next, benefits applications with limited annotated data from large publicly available images) and the creation of DL libraries facilitated further use of DL techniques for domain applications [21, 24]. Figure 3 describes stage of CNN. A literature study is thus needed to synthesize the available information, best practices, limits and possible solutions to DL-technologies in the field of plant physical composition.

Aside from performance enhancement, research has been performed to better understand the mechanism of CNNs. This leads to the development of approaches for explainable artificial intelligence, which aids in the construction of interpretable and inclusive machine learning models and the confident deployment of the models. This would be very beneficial for scholars attempting to comprehend the CNN process and enhance architectural design. The work also demonstrated that learnt features may be applied to other classifiers, implying that CNNs might learn broad representations of pictures rather than specialized categorization characteristics. Successive work advanced this path, developing a variety of gradient-based approaches for visualizing the importance/relevance of features in classification results.

For this, we identify and describe Agricultural Plant Leaf Disease Identification and Classification which detects the crop disease and alerts farmer before suffering from any big loss. Here, we are using the YOLO (You Only Look Once) algorithm [16,18,26] to detect diseased crops.

3. Related Work

In this topic we discussed with some the related work done by various researcher in this area using different tools and techniques for our reference study point of view.

Sachin C. et al. [8], define a methodology that will detect and classify three different green vegetables of different sizes. This method they involve the use of Tensor Flow, Dark flow which is a Tensor flow version of 'You Only Look Once' (YOLO) algorithm and OpenCV. To train the desired network, several various vegetable images were fed into the network. Before training, the images were pre-processed. They were pre-processed by drawing bounding boxes around the vegetable manually using OpenCV. YOLO is actually the main algorithm which is responsible for detection as well as classification. This method they provide a faster and smarter way to identify an object in the given image or video. Once the network is trained, the test input is passed into the network and when this is done, the output will display bounding boxes around the recognized vegetable and label it with its predicted class category with accuracy of 61.6 percent [8]. The first step in training is to gather the related images which are basically images of different types of cucumber, green apple and green capsicum images from internet sources. In order to get an efficient and high accuracy classification, train as many images. A total of one hundred images of each vegetable was taken from various internet sources, of which 60 images were used for training and the other 40 images used for testing. In this case, there are three classes, i.e. cucumber, green apple and green capsicum. The number of output channels of a layer is given by filters variable. This rectangular selector is used as shown to point out where the vegetable is present in each of the training images and a corresponding xml file is created which stores the top left coordinates and the bottom ride coordinates and the class to which the vegetable belongs. This xml files will be later fed inside the network to train the network. The entire images are divided into batches of size 16 images each and in each step of training the entire batch will go through the hidden layers and the weights will be updated correspondingly. Since the images contain 180 images, there are 11 batches and each epoch having 11 steps. To correctly detect and classify the presence of the vegetable in an image, a script is run that loads the trained YOLO model which in-turn loads the images that need to be classified. The script also identifies the coordinates and the class of the vegetable. A threshold of 0.15 is set which means that the model classifies only those bounding boxes that have confidence score of 0.15 or greater. The testing is done on three different vegetable images on different parameter. The images that has been used consists of 180 test images that comprises of mixed settings of all the three vegetables viz. it being horizontally placed or vertical or under complex backgrounds or as a bunch. For the images as test set, more than 50% of the images were detected appropriately and the vegetables were classified correctly and when a video was provided as an input to the model, 70% of the vegetables were properly detected and more than 70% of time the vegetables were correctly classified. The model was successful in detecting vegetables and classifying them in varying conditions and orientations which included vertical, vegetables in a bunch, and some of complex backgrounds were successfully detected. The model also gives a confidence score for each of its prediction and that prediction score was greater than 50% for almost all the images.

The authors in [9] have taken tomato plat leaf to identify disease using image processing techniques based on image segmentation, clustering, and open-source algorithms, thus all contributing to a system which is reliable, safe, and accurate for leaf disease identification with the specialization for Tomato Plants. In this DWT and GLMC feature extraction is used then segmentation process then CNN classifier for detection and classification. They have used gaussian filter with smoothened approach for enhancement of tomato leaf. They rectify the detection of tomato leaf disease for the prediction of affected and normal leaf. More number of samples need to be tested survey on different disease classification. The tomato leaf images associated with relevant diseases were taken for detection and classification for the proposed method. They received very optimum results. They stated that the method is compared with automated capturing systems using Deep Neural Networks (Alex-Net) and ANN technique and found high accuracy rate of their method due to a hierarchical feature extraction that maps the pixel values and evaluates the same with the trained images.

The authors in [10] used Zernike Moments (ZM) and Histogram of Oriented Gradients (HOG) method for classification of plant leaf images are proposed in this paper. After preprocessing, they compute the shape features of a leaf using Zernike Moments (ZM) and texture features using Histogram of Oriented Gradients (HOG) and then the Support Vector Machine (SVM) classifier are utilized for classification of plant leaf image and recognition. For the experiment purposes they first convert RGB image to binary image but before that each RGB image is converted into grayscale after which they calculate a threshold using Otsu's method and using this threshold level, convert the grayscale image to binary leaving the leaf image in white and background in black. Then eliminating the petiole is the next step. Some leaves have petioles so they have to be eliminated because they can distort the overall shape. For that the Distance Transform operator which applies only to binary images is used. It computes the Euclidean distance transform of the binary image. For each pixel in binary image, the distance transform assigns a number that is the distance between that pixel and the nearest nonzero pixel binary image. The authors have used Zernike Moments (ZM) on centered binary images and Histogram of Oriented Gradients (HOG) on cropped grayscale images to extract and calculate features of leaf image. Zernike Moments (ZM) help to extract features using the shape of leaf. The computation of Zernike moments from an input image consists of three steps: computation of radial polynomials, computation of Zernike basis functions, and computation of Zernike moments by projecting the image on to the basic functions. Histograms of Oriented Gradients (HOGs) are featured descriptors used in computer vision and image processing for the purpose of object detection. Finally, to test the method, they have first used Flavia images, then they have used the Swedish Leaves images and at last they combined both the Flavia and Swedish Leaves images. They have 47 species from both images, 32 species from Flavia images and 15 species from Swedish Leaves images. In Flavia images they used 50 samples per species. For each species they used 40 samples for training and 10 for testing. In Swedish Leaves images they have used 50 samples per species for training and 25 samples for testing and second, they have used 25 samples for training and 50 for testing to compare results with other methods. Using Zernike Moments (ZM) up to order 20 and Histogram of Oriented Gradients (HOG) as features, resulted in a high accuracy of 97.18% using only Flavia images, 98.13% for using only Swedish Leaves images and 97.65% for using both images.

The authors in [11] identify and use the problem of detection accuracy and in neural network approach support vector machine (SVM) is latest classifier of that approach. In their work, SVM has been implemented which contains two images; one is training images and train images. Firstly, original image is captured and then it is being used for processing. Secondly it gives us the black and background pixels of image segmented and also separate the hue part and saturation part of image. Thirdly detection of disease and diseased part of image is detected and healthy part is segmented from it.

Now, mask image is an image in which values are set of some of the pixel intensity values to zero and other values as non-zero. Now wherever the value of the pixel intensity is finding as zero in the mask image, then the intensity of the pixel of the resulting masked image will be going to set to its background value which is normally zero. In this section, experimental study and its results mainly will focus on SVM algorithm. This algorithm will first take an input image which is in RGB form. It detects the infected part of the disease. They also provide % of area in which diseases are present and give us the name of disease. As in the results of one image diseased area is 5.56%. This work provides accuracy which is better results. In this author has used PCA, SVM, Enhanced and neural network algorithm for getting better accuracy in which image segmentation and statistical analysis model are also used.

The authors in [12] use digital cameras and continuous improvement in computer vision domain, the automated techniques for detection of disease are highly in demands in precision agriculture, highly productive plant phenotype, smart greenhouse and much more. Working on an open reference images which has15200 images of crop leaves, a Residual Network (ResNet34) was trained to perform this task of classification. The illustrating the viability of the proposed model.

Overall, the process of training ResNet models on an open image provides a sound way towards crop disease detection using automated networks on an enormous global scale. The author has compared the ResNet algorithm result with SVM, decision tree, logistic regression and K-NN by performing accuracy and precision. Res-Net34CNN architecture is used for classification purpose in the proposed method. CNN is a special type of artificial neural network (ANN) which is customized for image processing.

The goal of a Conv-Net is to decrease the size of an image without losing crucial features that helps in solving the problem. By increasing the depth of the network, the accuracy tends to get saturated and then degrades quickly. This is known as the degradation problem.

Surprisingly, over-fitting is not the cause. The phenomenon of vanishing/exploding gradients in deep neural networks leads to this notorious degradation problem.

In the vanishing gradient problem, the gradients become infinitely smaller due to repeated multiplication during the back propagation step, resulting in negligible updates to the parameters. Exploding gradients are an issue. Here, gradients accumulate and lead to big updates to the parameters during training, preventing the model to learn from data. A residual block makes the use of skipping connections to address the degradation problem. Connections that skip one or more layers are Shortcut connections. ResNet34 architecture made from a plain-34 layer convolutional neural network by just introducing skip connections.

Arun Pandian J at.al.[13] used data augmentation increases the diversity of training more images for machine learning algorithms without collecting new image. In this work, augmented plant leaf disease images have developed using basic image manipulation and deep learning-based image augmentation techniques such as Image flipping, cropping, rotation, color transformation, PCA color augmentation, noise injection, Generative Adversarial Networks (GANs) and Neural Style Transfer (NST) techniques. Performance of the data augmentation techniques was studied using state-of-the-art transfer learning techniques, for instance, VGG16, Res-Net, and InceptionV3. An extensive simulation shows that the augmented images using GAN and NST techniques achieves better accuracy than the original images using a basic image manipulation based augmented images. The result outcome of transfer learning through data manipulation is 89%. Table 1 shows the analysis of various methods and sample images were used in Literature survey.

Table 1. Summary of sample reference works

No.	Title of reference reviews	Method	Sample Images
1.	Vegetable Classification Using You Only Look Once Algorithm [8]	You only look once, deep learning	Cucumber, green apple, green capsicum

2.	Tomato Leaf Disease Detection Using Deep Learning Techniques [9]	Convolutional neural network and segmentation.	Tomato leaf
3.	Plant Leaf Recognition Using Zernike Moments and Histogram of Oriented Gradients[10]	Histogram and linear support vector machine.	Flavia and Swedish leaves
4.	An Enhancement in Classifier Support Vector Machine to Improve Plant Disease Detection [11]	Enhanced Support vector machine	Different Plant leaf sample
5.	Res-Net based approach for Detection and Classification of Plant Leaf Diseases [12]	Resnet parameter with CNN	New Plant leaf disease
6.	Data Augmentation on Plant Leaf Disease Image Dataset Using Image Manipulation and Deep Learning Techniques [13]	Generative adversarial networks and neural style network	Plant Leaf disease

4. Problem Statement

There are many rural areas where agricultural farming the farmer couldn't decide or making decision or complex decision while working on large farm or agricultural, they need to be sure of when it need to be done in order to increase the yields and precision of the crop and they couldn't get accurate decision more error less yields improvement for protection of agriculture plants. Here suggestion needs to be accurate in order to earn more profit.

While the existing system has reached to certain reliability and accuracy in right algorithm and technologies, still need to get more accurate for more accuracy in agricultural plant or crop identification. Many research articles and publication have used different algorithm to identify and classify the plant leaf disease but when the problem is location where we want to detect the particular region and it implement the part it doesn't grab the correct one and accuracy get low. Out of that referred few methods and algorithm is included here.

5. Proposed Work

Thus, our main goal is to develop such a system using various latest computer-based tools like flask server and ResNet 50 image classification model based on artificial intelligence/ machine learning techniques. By using the model passes various sample of leaf images then getting information about classifications and various features of that plant leaf disease. Figure 1 shows our work approaches diagram system Architecture.

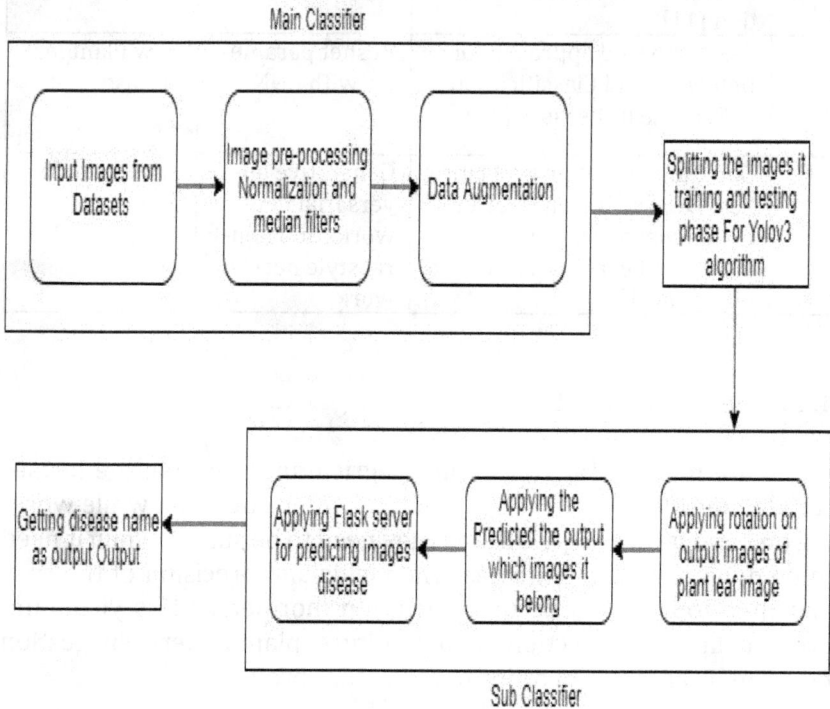

Figure 1. Proposed system diagram.

5.1 Main Classifier

The main parts are collecting different categories of plant leaf disease images. It is very difficult to collect lead images in only one season in agricultural plant, because in different seasons farmers grow different crops and find different disease on leaves. For our experiment for training purpose, existing sample images consisting of 15 features of three different leaves namely

- Potato leaf,
- Bell pepper leaf, and
- Tomato Leaf, were taken.

Also, collected nearly 1000 images of different plant leaf disease photographs. We found that following are the major categories of plant diseases in potato, bell pepper and tomato leaves:

1] Bell pepper Bacterial Spot
2] Bell pepper Healthy
3] Potato Early Blight
4] Potato Healthy
5] Potato Late Blight
6] Tomato Target Spot
7] Tomato Mosaic Virus
8] Tomato Yellow Leaf Curl Virus
9] Tomato Bacterial Spot
10] Tomato Early Blight
11] Tomato Healthy
12] Tomato Late Blight
13] Tomato Leaf Mold
14] Tomato Septoria Leaf Spot
15] Tomato Spider Mites Two Spotted Spider

Basic execution steps of proposed work.

5.1.1 Pre-Processing

Getting the preprocessing details about different images. In above listed of images given the reference name. In preprocessing data augmentation, normalization and median filter process is done to get more images sample for getting better accuracy.

5.1.2 Normalization

Image normalization is a common image processing technique for modifying the range of pixel intensity values. The word "normalization" refers to the process of converting an input image into a range of pixel values that are more familiar or normal to the senses.

We conduct a function that generates a normalization of an input picture in this project (grayscale or RGB). Then we comprehended a representation of the image's scale's range of values, which is represented between 0 and 255, and as a result, really dark images became clearer. The formula for linear normalizing of a digital picture is as follows:

Output_channel = 255 * (Input_channel - min) / (max-min)

If we're working with a grayscale image, we just need to use one channel to normalize it. If we're normalizing an RGB (3-channel) image, we'll need to apply the same criterion to each channel.

5.1.3 Median Blur Filter

The main disadvantage of the ordinary median filter is that it replaces the pixels under consideration with the window's median even if they are uncorrected (different than 0 or 255). This will degrade the image's overall visual quality. As a result, when it comes to maintaining important detail in a picture, the median filter generally outperforms the boxcar filtering approach [15].

It works as follows: The window sorted in ascending. After sorting, the median is the value in the center. As a result, the undamaged pixels are replaced by the window's median value. Sample window output as shown below, if the considered pixel is destroyed, the impulse noise will be removed in the same way [15].

$$
\begin{bmatrix} 46 & 64 & 82 \\ 255 & (255) & 52 \\ 64 & 64 & 82 \end{bmatrix} \longrightarrow \begin{bmatrix} 46 & 64 & 82 \\ 255 & (82) & 52 \\ 64 & 64 & 82 \end{bmatrix}
$$

The original pixel distribution, on the other hand, is not taken into account during the recovery process. In natural photographs, there is usually a substantial connection between neighboring pixel values.
In a natural picture, there are frequently substantial relationships between nearby pixel values. As a result, replacing noisy pixels with nearby values outside the median value is sometimes more accurate, it preserves edges while removing noise.

5.1.4 Image Augmentation

Image augmentation is techniques for transforming the image to clarify the image more clearly which cannot detect clearly. The augmentation technique can transform images like various operations on images, such as zoom, flip, rotation, brightness, sheer range [28].
This technique will create new image in a given input image to detect the algorithm more precisely with the help of image augmentation. The pre-processing filter is not enough for the images so the image augmentation is done.

5.1.5 Data Annotation

Data annotation is the process of reading the coordinates of images that are present in the various images. This data annotation is used for labelling the image. The best tools to annotate the images is LabelImg. This image is annotated reading the exact coordinate as in object detection data annotation is a main purpose.

This annotation will do two this one is coordinate of the images and also according to the image class will be define particular index of the images that they belong.

5.1.6 Feature Extraction

In feature extraction the parameter of Darknet-53 layer neural network will work as backbone of the YOLO V3 algorithm and the feature extraction will be compared to different algorithm. The different transfer learning techniques will be trained and be compared to the proposed model.

The You only Look algorithm v3 trained and the output images of the algorithm will be pushed to transfer learning algorithm to check to belong to particular class and then it passed to ResNet-50 for predicting the images values.

5.1.7 Detection

The detection or identification of the images is done using OpenCV library that will read weights of the trained image and will be tested and from the will we feed the images to get the accuracy of the model.

The detection is done to get to know which disease it belongs according to that the treatment is done. So, to identify the different diseases images this disease and flask server will help to get the output. The detection works as identification the different plant leaf disease showing which disease it is so it can be easier to understand.

5.2 Sub Classifier

In sub classifier the output obtained in main classifier is fetched from the main classifier and then rotation of main classifier is done new directory then prediction is done for predicting the output. This prediction is multi-stage classifier where the main-classifier output is passed to the sub-classifier and then to ResNet50 which is CNN transfer learning model it will classify the output label. Then through applying flask server the prediction can done through real-time deployment.

In the sub classifier the main classifier output works as a two-stage classifier for prediction the image of main classifier is first cropped and then it been rotated and then using prediction it will display disease output.

Then flask server is deployed for getting the output of the images from the local directory to predict the images. The flask server is done in Spyder IDE.

6. Discussion and Analysis of Outcomes

6.1 Pre-Pprocessing Sample Images

We have taken some image samples, which contains thousands of different plant leaf images having various diseases for training purpose only. In our algorithm we performed mainly 15 different categories of plant leaf disease for observation and experiment outcomes.

6.2 Applying Normalization

The normalization will be used in sub classifier which will be trained in ResNet50 CNN model and this normalization will be used for scaling down the images. When scaling down images, the value of the image will be from 0 to 255 range.

This scaling will also help with flask server where in flask server will take input from directory and then it will scale down from 0 to 1 to predict the images of which image is matched with disease and predict the output.

6.3 Applying Median Filter with Depth Density 3

Median filter is used for blurring the images for removing the noise from the original images and remove unwanted noisy image to get more accuracy. Median filter is used with depth density of 3 parameter to get litter blur images. In median filter depth density can be of value 1 and 5 for blurring the images. But for our experiment value 3 gives better accuracy and also 5 depth density get more blur and is not visible clearly.

6.4 Data Augmentation with Aifferent Parameters

In data augmentation we create different new samples for training the images to get trained images of different looks from same sample datasets. We have taken parameter of rotation range 40 which will rotate at 40-degree, width and height range of 0.2 and sheer range of 0.2, zoom range of 0.2 and horizontal flip for flipping images for creating new samples which will help for experiment.

This parameter is first observed and used using different value in which this value gives better results for identifying the image because in some parameter value where value increases, the image gets shifted more and doesn't recognize it. So, for this experiment we have taken theses parameter values.

The Pre-processed image are being trained in yolov3 for getting trained for co-ordinate in Googlecolab for faster GPU process and after it's been trained then passed or being applied for getting confidence using threshold and NMS confidence for output. So, it can complete the main-classifier.

According to the base paper the images are note been pre-processed and only been trained for yolov3 algorithm which gives less accuracy.

6.5 Bounding Images with Non-max Suppression and Cropping images

According to this operation will get actual location of diseases found perfectly and make it more clear classification.

6.6 Applying Rotation on Main Classifier Results

During the classification of a leaf based on plant it was observed that in spite of augmenting the data over a wide range of rotation the model performed better when an image is in upright position than when rotated over a large range. This gives a better prediction of results.

6.8 Results in Flask Server

Use of flask server deployment for predicting plant leaf disease is done. In the flask server the images are picked from directory randomly and then the prediction is done in GPU to predict the disease name.

6.9 Sample Images Rested Results

Step 1: Output result of median filters

Step 2: Applying Data Augmentation results

Step 3: Accuracy results of input leaf images

Table 2. Statistics parameter performance of algorithm

Description	Parameters
Number of images tested on plant leaf disease	250
Average confidence score	90%
Percentage of images with true positive	75%
Percentage of images with false positive	30%
Percentage of images with confidence scoregreater than 50%	98%
Number of images with confidence greater than95%	40%

Different parameters for measures the performance of algorithm shown in Table 2.

Table 3. Plant disease leaf identification and classification

Parameters	Bell pep-per Leaf	Potato Leaf	Tomato Leaf
Successful detectionand classification	90%	87%	81%
False positive percentage	40%	50%	60%
Undetected plant leaf disease	20%	20%	40%

Applied identification and classification technique to find the Plant disease on leaf shown in Table 3.

Table 4. Plant disease leaf identification and classification

Plant Leaf	Number of training images	Number of testing im-ages	Diseases	Accu-racy
Bell pepper Leaf	450	200	Pepper Early Blight	97%
Potato leaf	580	320	Potato Late Blight	89%
Tomato Leaf	900	487	Tomato Mold leaf	91%

Table 4 shows our experimental results of identification of disease with accuracy of our proposed system.

7. Conclusion

Many deep learning models are developed on the base of CNN for detection and identification and classification problem by many researchers. A feedback system offers suitable knowledge, treatments, disease prevention and control measures that result in better crop yields can be coupled with the suggested structure.

Our experiment for training purposes was done first on existing images and then after taking final testing on three different leaves like Potato leaf, Tomato Leaf, Bell pepper, we have done experiments on our proposed system and achieved the best result for in identification and classification of leaf disease.

We achieved highest accuracy in bell pepper leaf detection of disease 97 percent. The drawback of this model it will take lot of time to train data augmentation image for algorithms but it will definitely be helpful for agriculture related researches. Additionally, the flask server can also be useful for real time to get the disease name and details about disease.

Future Work

In further work we would collect real time images from different season of plant leaf disease.

The trained models are suitable for early and automated detection of plant disease. Preventive measures may be taken more quickly. The utilization of state-of-the-art technologies, such as smartphones, cameras and robotic platforms, might assist to identify diseases early and automatically in diverse crops.

References

1. K. Fukushima (1980) Neocognitron: a self-organizing neural network model for a mechanism of pattern recognition unaffected by shift in position, *Biological Cybernetics,* 36(4), 193–202.
2. Krizhevsky A., Sutskever, I., Hinton, G. E. (2012) Imagenet classification with deep convolutional neural networks, *Advances in Neural Information Processing Systems*, Curran Associates, Inc., Stateline, NV, USA.
3. Davis, S. L., Dukes, M. D. (2010). Irrigation scheduling performance by evapotranspiration-based controllers. *Agricultural Water Management*, 98 (1), 19-28.
4. Gutirrez, J., Francisco, J., Villa-Medina, J. F., Nieto-Garibay, A., Porta-Gándara, M. Á. (2014). Automated Irrigation System Using a Wireless Sensor Network and GPRS module, *IEEE Transactions on Instrumentation and Measurement,* 63(1), 166-176.
5. Kaur, S., Pandey, S., Goel, S. (2018). Semi-automatic leaf disease detection and classification system for soybean culture, *IET Image Processing*, 12(6), 1038-1048.
6. Venugoban, K., Ramanan, A. (2014). Image classification of paddy field insect pest using gradient-based features, *International Journal of Machine Learning and Computing (IJMLC)*, 4 (1), 1-5.
7. Anami, B. S, Malvade, N. N., Palaiah, S. (2020). Deep learning approach for recognition and classification of yield affecting paddy

crop stresses using field images. *Artificial Intelligence in Agriculture* 4, 12-20.

8. Garcia-Garcia A., Orts-Escolano, S., Oprea, S., Villena-Martinez, V., Martinez-Gonzalez, P., Garcia-Rodriguez, J., (2018) A survey on deep learning techniques for image and video semantic segmentation, *Applied Soft Computing*, 70,41–65.

9. S. C, N. Manasa, V. Sharma, N. K. A. A. (2019) Vegetable Classification Using You Only Look Once Algorithm, *International Conference on Cutting-edge Technologies in Engineering (ICon-CuTE)*, pp. 101-107, doi: 10.1109/ICon-CuTE47290.2019.8991457.

10. Ashok, S., Kishore, G. Rajesh, V., Suchitra, S., Sophia, S. G. G., Pavithra, B. (2020) Tomato Leaf Disease Detection Using Deep Learning Techniques, 2020 *5th International Conference on Communication and Electronics Systems (ICCES)*, pp. 979-983, doi: 10.1109/IC-CES48766.2020.9137986.

11. Tsolakidis D.G., Kosmopoulos D.I., Papadourakis G. (2014) Plant Leaf Recognition Using Zernike Moments and Histogram of Oriented Gradients. *In: Likas A., Blekas K., Kalles D. (eds) Artificial Intelligence: Methods and Applications. SETN 2014. Lecture Notes in Computer Science,* vol 8445. Springer, Cham. https://doi.org/10.1007/978-3-319-07064-3_33.

12. Rajleen Kaur, Dr. Sandeep Singh Kang (2015) An enhancement in classifier support vector machine to improve plant disease detection, *IEEE 3rd International Conference on MOOCs, Innovation and Technology in Education (MITE)*, 2015, pp. 135-140, doi: 10.1109/MITE.2015.7375303.

13. Kumar, V., Arora, H., Harsh, Sisodia, J. (2020) ResNet-based approach for Detection and Classification of Plant Leaf Diseases, *International Conference on Electronics and Sustainable Communication Systems (ICESC)*, pp. 495-502, doi: 10.1109/ICESC48915.2020.9155585.

14. Arun Pandian, J. A., Geetharamani, B. Annette. (2019) Data Augmentation on Plant Leaf Disease Image Dataset Using Image Manipulation and Deep Learning Techniques, *IEEE 9th International Conference on Advanced Computing (IACC)*, 2019, pp. 199-204, doi: 10.1109/IACC48062.2019.8971580.

15. M. Pagliai, N. Vignozzi, S. Pellegrini, (2004) Soil structure and the effect of management practices, *Soil and Tillage Research*, Vol. 79, No. 2, pp. 131-143.

16. Gavhale, K. R. Gawande, U., Hajari, K. O. (2014) Unhealthy region of citrus leaf detection using image processing techniques, *International Conference for Convergence for Technology-2014*, pp. 1-6, doi: 10.1109/I2CT.2014.7092035.

17. Morbekar, A., Parihar, A., Jadhav, R. (2020) Crop Disease Detection Using YOLO, *International Conference for Emerging Technology*

(INCET), 2020, pp. 1-5, doi: 10.1109/INCET49848.2020.9153986

18. Mekala, M. S.; Viswanathan, P. (2017) A novel technology for smart agriculture based on iot with cloud computing. In Proceedings of the International Conference on IoT in Social, Mobile, Analytics and Cloud (I-SMAC), pp. 75–82.

19. Ponnusamy, V., Coumaran, A., Shunmugam, A. S., Rajaram, K., Senthilvelavan, S. (2020) Smart Glass: Real-Time Leaf Disease Detection using YOLO Transfer Learning, *International Conference on Communication and Signal Processing (ICCSP)*, pp. 1150-1154, doi: 10.1109/ICCSP48568.2020.9182146.

20. Agarwal, M., Kaliyar, R. K., Singal, G., Gupta, S. K. (2019) FCNN-LDA: A Faster Convolution Neural Network model for Leaf Disease identification on Apple's leaf dataset,*12th International Conference on Information & Communication Technology and System (ICTS)*, pp. 246-251, doi: 10.1109/ICTS.2019.8850964.

21. Vaishali, S., Suraj, S., Vignesh, G., Dhivya, S., & Udhayakumar, S. (2017). Mobile integrated smart irrigation management and monitoring system using IOT. . *International Conference on Communication and Signal Processing (ICCSP)*, pp. 2164-2167.

22. R. G. de Luna, E. P. Dadios, A. A. Bandala. (2018) Automated Image Capturing System for Deep Learning-based Tomato Plant Leaf Disease Detection and Recognition, *TENCON 2018 - 2018 IEEE Region 10 Conference*, pp. 1414-1419, doi: 10.1109/TENCON.2018.8650088.

23. Sardogan, M., Tuncer, A., Ozen, A.(2018) Plant Leaf Disease Detection and Classification Based on CNN with LVQ Algorithm, *3rd International Conference on Computer Science and Engineering (UBMK)*, pp. 382-385, doi: 10.1109/UBMK.2018.8566635.

24. Ganesh, P., Volle, K., Burks, T. F., Mehta, S. S. (2019) Deep Orange: Mask R-CNN based Orange Detection and Segmentation, IFAC-PapersOnLine, 52(30),70-75. https://doi.org/10.1016/j.ifacol.2019.12.499.

25. P. R. Rothe, R. V. Kshirsagar (2015) Cotton leaf disease identification using pattern recognition techniques, *International Conference on Pervasive Computing (ICPC)*, pp. 1-6, doi: 10.1109/PERVASIVE.2015.7086983.

26. Liakos, K. G., Busato, P., Moshou, D., Pearson, S., Bochtis, D. (2018). Machine Learning in Agriculture: A Review. *Sensors*, 18(8), 2674, 1-29.

27. Jashraj Karnik, Anil Suthar (2020) A survey of plant leaf disease identification using deep learning techniques, *Dogo Rangsang Research Journal*, ISSN 2347-7180, 10(12), 91-95.

Machine Learning: The Future of Agriculture

Anuja Fole and Nilay Khare
Department of Computer Science and Engineering,
Maulana Azad National Institute of Technology, Bhopal, India

Abstract: Machine learning is gaining more popularity nowadays due to the increase in the generation of a huge amount of data every day by a single individual, lots of mathematics bases are available for calculations and many new technologies like big data and high computing processors are making it possible to process those data. So, machine learning techniques can also be implemented along with agro-technologies to improve productivity in agriculture. The chapter includes an overall review of how these algorithms have been applied to various divisions in agriculture management. The chapter talks about crop management, soil management, water management, livestock management and other such things related to agriculture. The review in this chapter will help in understanding how different machine learning algorithms can help farmers in making the best decision for their crops and lands. Also, it can help the researchers to identify how and which applied algorithms can be optimized by tuning their parameters.

Introduction

The agricultural segment plays a significant role everywhere as it is a major source of production of food. With the continuous development in the world, agricultural land is decreasing year by year in many places. Different technologies have already been proposed and introduced in agriculture for helping the farmers with agricultural work, however, they are not yet widely implemented and still are out of reach of the farmers.

Not only the decrease in agricultural land is an issue but also due to global warming the climatic conditions are changing day by day. This will be impacting agriculture more in near future. It is the time that different useful technologies which can help farmers in their agricultural work be made available for them so that they try to implement it.

For example, using technology like internet of things (IoT) the data from the agricultural land and atmosphere can be monitored. This data can be stored in a central repository for further processing. This data can help the researchers to provide better learning to solve any agriculture-related issues.

IoT can be used in more different applications in agriculture as well like for automating irrigation. Similarly, other different technologies can be used in agriculture like machine learning which can be helpful for agriculture. We will first understand what machine learning is and then understand its application in agriculture and how it can help.

An overview of Machine Learning

Machine learning is a device's ability to know itself, without being coded directly.

A proper definition can be "A computer program is said to learn from experience E with respect to some class of tasks T and performance measure P, if its performance at tasks in T, as measured by P, improves with experience E." [3].

The data that is provided as input is a set containing examples, where each example is further illustrated using various attributes called features. The types of data a feature can have are categorical, numeric, or binary. The performance metric is the measure of performance that aids in improving the model with experience.

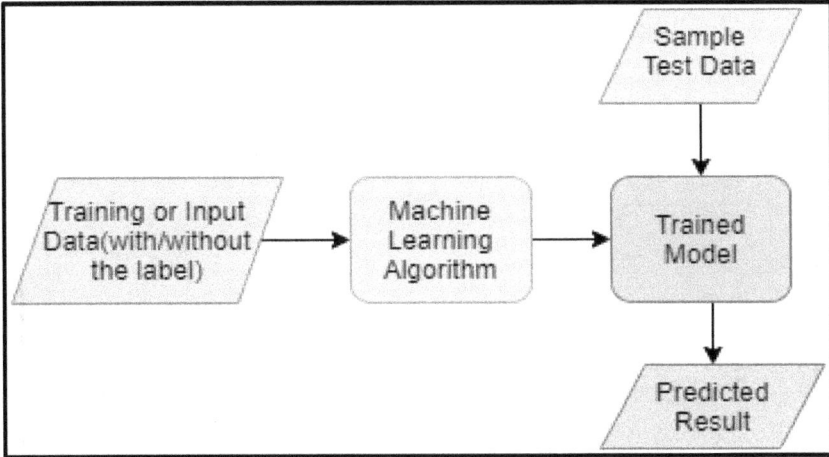

Figure 1. The general working of the machine learning models

In general, out of the total sample inputs, nearly 80% of the data is used for training and once the model is ready it is then tested with the rest of the 20% data. Testing is done to see how accurate our trained model is predicting the results.

Categories of Machine Learning

There are usually 3 categories of machine learning, they are defined as 1) supervised (or guided) learning, 2) unsupervised (or unguided) learning, and 3) reinforcement learning. In the first category (supervised) of learning, the training samples also contain the label or the target output.

This is further divided into classification and regression. In the second category of (unsupervised) learning, the unlabelled data is fed to the model and the model itself has to decide how to proceed with the data. In reinforcement learning, we have an actor and environment. An actor performs the task and gets a reward as feedback and using this it tries to improve its behaviour and acts accordingly. Figure 2 shows the categorization of models.

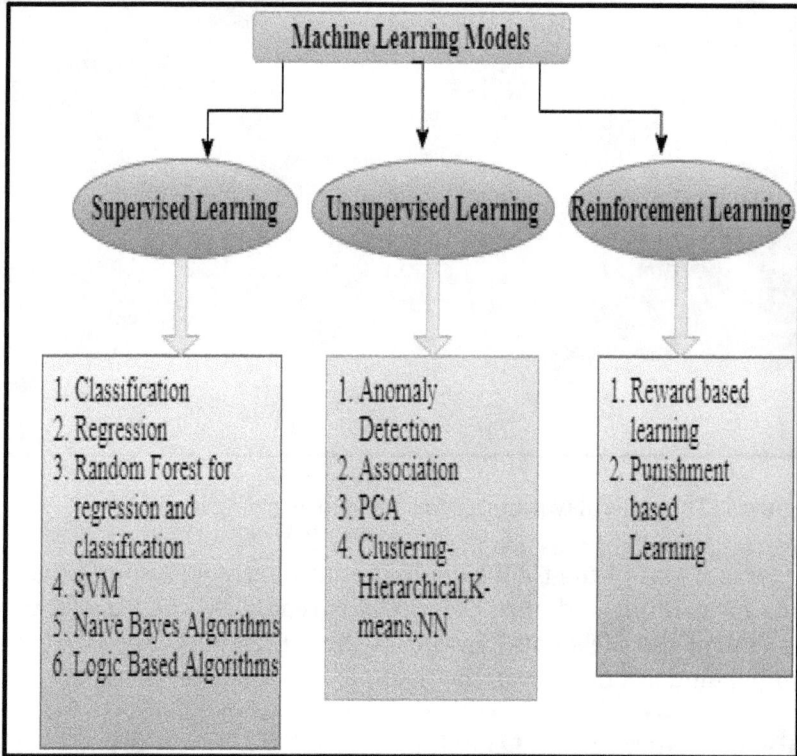

Figure 2. Categories of machine learning models.

Feature Selection and Extraction

Selection help in plummeting the number of input features that can be actually utilised for further processing by selecting a set of features which is optimal and helps to preserve the original information as much as possible.

Learning Models

Implementation of models of learning in ML is limited to those incorporated in the works mentioned in this review.

1. *Regression Models*:
   It is a supervised model of learning where the prediction output variable values depend on known input variables. We

have different types of algorithms including linear regression and logistic regression, and step by step regression [29]. More complicated regression algorithms such as ordinary least-square regression [30], multivariate adaptive regression splines [31], multiple linear regression cubist [32], and locally estimated scatterplot smoothing [33] were also introduced.

2. *Clustering Models:*
   It is one of the unsupervised learning models. Clustering is the process of splitting the data points or the population into several groups, in such a way that data points in the same groups are more comparable to other data points in the same category than others. Simply put, the goal is to segregate and assign groups with similar traits into clusters. There are various clustering techniques like k-means, hierarchical, and expectation-maximization techniques at present.

3. *Bayesian Models:*
   A supervised learning model that is helpful in solving regression and classification problems. Its foundation is based on probabilistic inferences. Different models like naïve Bayes, multinomial naïve Bayes, Bayesian belief network, Gaussian naïve Bayes, and a mixture of Gaussians are currently used.

4. *Decision Trees Models:*
   Decision Trees are formulated in the tree-shape structure. DT can be used as a regression and classification model. It contains a root, internal nodes, and leaf nodes. Each of these nodes is formed using the features in the dataset. Every branch represents the output. Leaf node represents the final or predicted outcome. To predict the result for test sample one must traverse from root to leaf node via internal nodes and taking the decisions on each node.

5. *ANN:*
   Artificial Neural Network takes its inspiration from the functionality of the human brain. It is the emulation of most complex tasks like decision making, pattern generation, learning, and recognition, storing, and processing information that our brain performs. The human brain is a composition of billions of neurons where the information after being processed at

one is passed to another. Similar thing is being represented using interconnected neurons arranged in multiple layers. We can define the different layers as:

1. The input layer where each node/neuron represents each feature and from where the data is fed.
2. The hidden layer: Number of hidden layers should be at least one and is responsible for the actual learning function.
3. The performance layer or output layer lets you map the result.

These supervised learning models can typically be intended for working out classification and regression problems.

ANN uses some of the common learning algorithms like the perceptron algorithm, radial basis function network, back-propagation, and resilient back-propagation. Learning algorithms like an extreme learning machine, multilevel perceptron, autoencoders, adaptive-neuro FIS (fuzzy inference systems), supervised Kohonen Networks, generalized regression neural network, ensemble averaging, XY-fusion, self-adaptive evolutionary extreme learning machine, and counter propagation are the currently ANN-based algorithms in use.

Deep learning is a new, innovative technique for computer vision and analysis of data, with amazing results and huge potential. Deep Learning is an extension of ML, where models are more complex, as well as there, are functions that transform the data which can be maintained in a hierarchical representation across multiple levels of abstraction. Deep learning models have the ability to automatically pull out the important features from the raw data, which we can call as a feature learning function. Deep learning models are derived from ANN in which they can have several hidden layers, containing additional or fewer neurons between the input and output layers. Depending on the network's architecture these models can either be unsupervised, semi-supervised, or supervised. The most commonly used DL models are Long Short-Term Memory (LSTM), Convolutional Neural Network (CNN), and Recurrent Neural Network (RNN). When dealing with images the model that takes all the credit is Convolutional Neural Network (CNN), where its major task is extracting important features from images. More on CNN can be found here [34].

6. *Support Vector Machine:*
   SVM is a supervised model of learning, that can be applied on problems of both the classes that is classification and regression. The algorithm's main objective is, that given a N-dimensional space it figures out the hyperplane in order to separately identify the data points. These hyperplanes are the decision boundary that represents the points present on either side belongs to different classes. It is not always the case that the available data would be linearly separable. So, for non-linearly separable data it relies on a technique called "kernel trick". Here the input data is transformed into higher dimensional space and then finds the optimum boundary in the new dimensional space. Variations of SVM that are mostly used are support vector regression [35], least-squares SVM [36], and successive projection algorithm-SVM [37].

7. *Ensemble Learning:*
   Ensembling is a technique of integrating the effects of a variety of individual learners to boost a model's predictive capacity and stability. These learners would differ on various parameters like the size of the population, use of hypothesis, different modelling techniques, different initial seed, etc. The main aim of this technique is that is one individual learner is stronger enough in producing accurate results so collecting such models would in any case going to enhance the system's efficiency and accuracy. After collecting the results different voting and stacking algorithms are applied to reach the conclusion. These approaches can be distributed into two sets. 1) sequential in which the generation of base learners is done sequentially as in the case of the Adaboost technique. This is used where the learners are dependent on each other.2) parallel: generating the base learner parallelly. An example is Random Forest. It requires less time as compared to sequential. Different ensemble models used are bagging, boosting, gradient boosting, stacking [38].

Literature Review

Agriculture has seen significant shifts in its complex parameters over the past few years. Some of them are listed below:

- Change in climatic conditions
- Degradation in soil quality
- Excessive use of very harmful pesticides
- Increase in different types of plant diseases
- Poor storage management
- Lack of knowledge of proper irrigation

Since the early '80s, computers and technology have shown keen interest in the above area and since then many proposed systems and suggestions came up for improvement in agriculture.

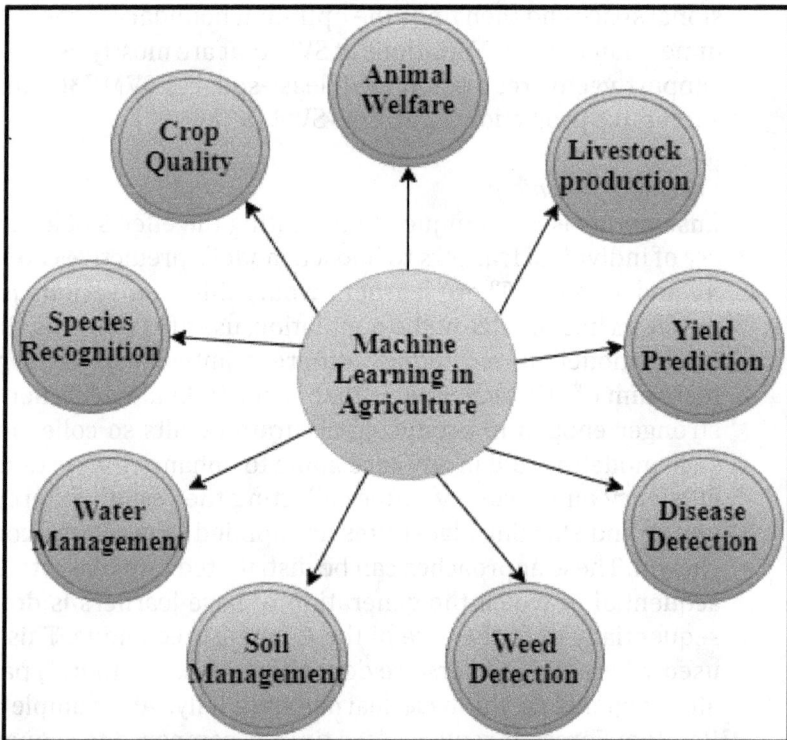

Figure 3. Areas of Agriculture discussed as review literature.

Various work under different agricultural issues have been discussed below:

1) *Choosing the Right Seed for the Area*:
   Most farmers are unaware as to which crop would be best suitable for their land. They only observe their neighbours for guidance. This leads to poor production on their land as the soil contents might not be for the grown crop. Machine Learning here can help to suggest the best crop according to the soil content.

2) *Plant Disease Detection:*
   With new diseases impacting plants every year, they either go unnoticed or being wrongly diagnosed. In most cases in India farmers have to consult expert advice on the disease which can be very costly and time-consuming. This results in damaging the crops and in turn reduction in yield. To tackle this problem different classification machine learning algorithms have been applied for correct disease detection. Here are the list of authors and the brief description of their work on CNN, K-NN, ANN, SVM, Fuzzy C-Mean classifiers.

   A) *Fuzzy C-Mean:*
      The author in [5] implemented the Fuzzy C-Means algorithm in order to recognize the existence of Rust disease on wheat leaves. The model was tested on the dataset containing the healthy and unhealthy images of the wheat leaves. It was able to segregate the two types of leave with 88% accuracy. It could recognize the leaf disease with an attainable accuracy of 56%.

   B) *Support Vector Machine Classifier:*
      The author in [6] presented the model for detecting diseases in various citrus crops like lime, grapefruit, lemon, and orange leaves infected by diseases anthracnose and canker. The trained model was able to produce 95% accuracy results. Powdery Mildew and Downy Mildew diseases in Grape plant [7] were being detected with an average accuracy of 88.89%. K -means clustering also used here for segmenting the diseased region and later extracted the colour and texture features. [8] SVM model

along with K-means clustering for feature extraction was able to detect palm oil leaf diseases Anthracnose with 95% accuracy and Chimaera with 97% accuracy. With 300 images of a potato plant disease called Late blight and Early blight, that were publicly available, the model [9] got the accuracy of 95%. Features of LAB and HIS colour model along with classifier helped to detect diseases of grape leaf-like Leaf Blight, Black Rot, and Esca [10].

In [11] the researchers built an SVM classifier to detect 2 most widely spread diseases of tea leaf specifically brown blight and algal leaf disease but with eleven features that are three less than the models in the past researches without much affecting the success rate. The model was able to maintain accuracy of more than 90%. [12] presented a model wherein K-means are used to separate healthy soybean leaves from unhealthy ones and then applying support vector machine to classify those diseases into categories named Septoria leaf blight, downy mildew, and frog eye.

C) *Artificial Neural Network:*
[13] proposed and evaluated a framework that detects plant leaf/stem disease using a feed-forward backpropagation algorithm. The model was tested using cottony mold, early scorch, late scorch, and tiny whiteness diseases and achieved an accuracy rate of around 93%.[14] developed a device that was able to detect the fungal diseases(powdery mildew and downy mildew) of cucumber plant leaves using image processing algorithms and ANN.[15] with a 90% accuracy model was able to recognize diseases like fruit spot, bacterial blight, fruit rot and leaf spot present in pomegranate plant using the backpropagation (BP) algorithm.[16]using conversion from RGB image to HSV and the applying backpropagation algorithm the model was able to recognize Cercospora(spot on the leaf) a disease found in groundnut plant with an accuracy of 97.41%.

D) *Deep Learning- Convolutional Neural Network:*

[17] proposed a plant disease recognition model using deep CNN which successfully classified diseases of 13 different types. The results were generated with accuracy in between 91% to 98% for different classes of tests with an average of 96.3%. The model was trained using 30880 images containing augmented images as well and was tested using 2589 images.

[18] Using over 54306 healthy and diseased leaves images from the public dataset, the researchers trained the deep CNN model categorised reported 14 types of crops and 26 leaf diseases and reached 99.35 per cent accuracy. They implemented two architectures one is Alexnet, which is trained from scratch with an 80-20 ratio of data, and achieved 85.53% while the other is GoogLeNet a pretrained network where only the last layer can be modified and achieved 99.34% of accuracy. [19] designed a model based on LeNet architecture for the classification of soybean plant disease. It was trained and tested using 12673 sample data having four classes achieving an accuracy rate of 99.32%. [20] deep learning model was developed using 87848 images of leaves from the open database having a combination of (plant, disease) as 58 different classes. The model was tested using 17548 unseen images and resulted in 99.53% accuracy.

E) *K-Nearest Neighbours:*
[21] trained a K-NN classifier model using two types of classes of sugarcane leaves: Normal and scorch. Feature extraction and image conversion steps were applied before training the model. It achieved an overall 95% accuracy. Deeper neural network (DNN) models or algorithms can be applied to improve the accuracy rate and also knowing the amount of infected area. [22] proposed a model that estimated the disease stage and detected the disease in cotton leaves with 82.5% accuracy over 40 images. [23] presented a model that used the GLCM algorithm to extract textual features and KNN for multiclass classification. It also reported the percentage of the in-

fected area over the leaf. The model showed a 10% performance increase of the existing model that used the SVM classifier and classified into two classes only.

All the classification techniques for plant diseases detection discussed above can be summarized as:

Table 1. Classification techniques for plant diseases

Classifier	Plant Dataset	Number of classes for diseases	Model Accuracy
SVM	Citrus Fruits [6]	2	Model had 95% accuracy.
	Grape [7]	2	Achieved average accuracy of 88.89%.
	Oil palm [8]	2	97% accuracy for one disease and 95% for another.
	Potato [9]	2	Reached 90% accuracy.
	Tea [11]	3	Achieved 95% accuracy.
	Soybean [12]	3	Approximately 90% accuracy.
ANN	Mix leaves data [13]	5	Around 93% accuracy shown.
	Cucumber [14]	2	Tried to improve the existing accuracy.
	Pomegranate [15]	4	90% accuracy achieved.
	Groundnut [16]	4	Got 97.41% accuracy.
KNN	Sugarcane [21]	1	95% accuracy model.
	Cotton [22]	1	Reached 82.5% accuracy.
Fuzzy	Wheat [5]	1	Disease recognition had 56% accuracy and detection had 88%.

CNN	Cherry, Peach, Apple, Pear and Grapevine [17]	13	96.3% average accuracy achieved.
	14 different Crops [18]	26	Managed to achieve 99.35% accuracy.
	Soybean [19]	3	99.32% accuracy achieved.
	25 plants [20]	58	Reached to a accuracy of 99.53%.

Apart from above machine learning algorithms, in [64] the author has used an optimization algorithm called PSO (particle swarm optimization) in order to detect soyabean leaf disease. The best part of this algorithm is that it does not require any before hand information about the count of segments, of which are rather required in the other existing models. Also, the model proved to require very less efforts in terms of computations. The model could attain an improved accuracy of 98.0%.

3) *Prediction of Crop Yield:*
This is a tougher task as it involves a great analysis of soil quality, weather conditions, rain predictions, geographical area, market, and pricing of the items. But nowadays the weather conditions are extremely unpredictable. These are highly impacting every stage of the plant growth cycle. Sometimes even the experts fail to correctly predict the yield even after studying all the parameters. Also, the use of the number of pesticides, weedicides (needed for getting rid of weed which is a plant which grows along crops where it is not required.), and fertilizers impacts the overall yield from the field. Complex Machine learning algorithms can help the experts to extract maximum information from the parameters in order to predict close to the exact yield. The method relies on different factors, such as cost of production, quality of the commodity and policies of government. Most researchers used statistical methods or ML techniques to study crop yield rate prediction, weather forecasting, classification of soil, and classification of field for agricultural planning. When there

are additional choices for crop plantation at a time on a limited availability of land resource, then picking a crop is a challenge. CSM (crop selection method) is often used to resolve the crop picking problem, to maximize the crop's net production rate over the period, and to achieve full country economic growth afterwards. The approach proposed will boost crop yield levels net. Below are researches done on the prediction of crop yield:

[24] Ghosh and Koley managed to train Back Propagation Networks (BPN) with the nutrient-available growing conditions of reference crops and their ability to supply nutrients from their own reserves and, in both cases, through external crop production applications, BPN led to the discovery and recommendation of the correct correlation percentage between those properties. A collection of test data was used to determine the efficacy of the Back Propagation Neural Network model and their results showed that artificial neural networks with the hidden layer having certain numbers of neurons performed better than multivariate regression in predicting soil properties.

Okori and Obua [25] showed which crop can be chosen in case of frequently occurring disasters like famine. They used various classifiers like SVM, KNN, Naïve Bayes, and decision trees and compared their performances. Out of which KNN has lower error rates than others. Its accuracy can be improved by training the model with more number of good quality images. [26] used the climatic variables: maximum temperature, potential evapotranspiration, minimum temperature, soil moisture, precipitation, and cultivated land to predict the maize production and used the data from the past 27 years for training the ANN model. The maximum accuracy for one of the region data reached 86% [27]. The parameters considered for this study were for the Kharif period (June to November), from a year 1998 to a year 2002, minimum temperature, precipitation, maximum temperature, average temperature, and evapotranspiration of reference crop, position, production and yields. The data set was analysed with the device WEKA. It has developed a Multilayer Neural Perceptron Network. The data was validated using a cross-validation approach. The results showed 97.5 percent accuracy with 96.3 sensitivity and 98.1 specificities. [28] shows the contrast between two recognized predictive systems for expecting wheat yields that are 1) multiple linear regression (MLR) and 2) artificial neural networks (ANN). Researchers looked at a variety of parameters such as precipitation level, soil evaporation, crop biomass, extractable soil water (ESW), transpiration, and the added nitrogen fertilizer applied. They found that more than MLR and D-ANN models, the Custom Artificial Neural Networks (C-ANN) display performed better on the test dataset with a lower rate forecast error and a higher R2 measurement.

[62] developed a method to map the carrot yield by applying random forest with gini index as parameter on the collection of spectral images from satellite and using georeferenced yield sampling. They have divided the complete dataset set into two parts: training and test sets. They evaluated the complete system on parameters like coefficient of determination (R-square), root mean square error (RMSE) and mean absolute error (MAE) and where able to achieve 0.82,2.64 and 1.74 respectively.

[63] aimed at developing a deep convolutional neural network model that could automatically help to recognise and classify different abiotic and biotic paddy crop stresses given the field images as input. The input dataset contained 30,000 site images, wherein the site had 5 varieties of paddy crop including 12 categories of stress(normal/healthy). With the help of pre-trained VGG-16 CNN model they where able to reach 92.89% as an accuracy.

Authors in [65] did a comparative analysis on ML and DL models in order to obtain the yield prediction and variation in growth of two different plants (tomato and Ficus bejamina) in greenhouse surrounding. They used deep recurrent neural network along with LSTM as their deep learning model versus the random forest regressor and support vector regressor as the machine learning models. Mean square error, root

mean square error and mean absolute error   were the evaluation parameters used for the comparison which showed
that RNN+LSTM outperformed the other models.

4) *Weed Detection:*
Detecting and controlling weeds is another big issue in agriculture. Weed is a plant which grows along crops where it is
not required. Most producers point to weeds as perhaps the
most severe crop production threat. Precise weed identification is of great importance for sustainable farming, as weeds
are hard to classify and they differentiate against crops. Again
yet, ML algorithms could lead to precise recognition and identification of low-cost weeds in combination with sensors,
without ecological problems and side effects. ML for unwanted plant detection may allow tools and robots to be established to kill weeds, which minimizes the necessity of
herbicides. Two studies have been reported on ML techniques for wild plant detection problems in farming. In the
study [39], the authors suggested an innovative approach for
identifying Silybum marianum built on counter-propagation
(CP)-ANN and multispectral pictures taken by unmanned air-
craft systems (UAS), an unwanted plant that is difficult to get
rid of and that leads to significant losses in crop yield. They
achieved 98.87 percent accuracy here, and observed features
such as green, red, and NIR spectral bands and the texture
sheet. In the next study [40], the authors introduced a new
way for the identification of crop and weed populations,
based on ML practices and hyperspectral images. More precisely, the authors constructed an active learning method to
Identify maize (Zea mays) as crop types, and Sinapis arvensis,
Cirsium arvense, Ranunculus repens, Stellaria media, Taraxa-
cum officinale, Polygonum persicaria, Urtica dioica, Poa an-
nua, Oxalis Europaea, and Medicago lupulina as unwanted
plant species. The basic goal was to better define and classify
certain animals for environmental and economic purposes.
Another study [41] the authors established an SVN-based
method of weed detection in grassland cropping. They took
photographic camera pictures of grass and wild plant as input
achieved different accuracy for different classification like
97.9% for rumex classification, 94.65 % for classification of

Urtica, and 95.1% for mixed weather conditions and mixed weeds.

Authors in [66] implemented the various available deep convolutional neural network in order to detect weeds namely dandelion, ground ivy and spotted spurge that are growing in ryegrass. These models were trained on a dataset consisting of 15,486 (having no target weed) and 17600 (having target weed) images. They implemented VGGNet with F1 score (>=0.9278) and high recall (>=0.9952), AlexNet with F1 score between 0.8437 to 0.9418, DetectNet with F1 score (>=0.9843) and GoogleNet didn't seem to detect weeds so well and kept getting lower precision values. Through their study they were able to come up with the fact that VGGNet and DetectNet would do wonders in this field. [67], authors here have focused on the pain area of manually labelling the data at the level of pixel. In order to accelerate the process, they have implemented a two-step method, wherein first they used maximum likelihood classification to segment the foreground and background and in second step they labelled the pixels of weeds manually. This labelled data was used in training the semantic segmentation models, that classify background pixel and crop as one segment and rest weed as another. Two different deep learning architecture like SegNet and UNET along with VGG16 and ResNet-50 was used for the experiment on canola field images. Their results showed that SegNet with ResNet-50 performed better than other with mean intersection over union (MIOU) as 0.8288 and frequency weighted IUO value as 0.9869.

5) *Quality of Crops:*
Studies formed for the understanding of features relevant to quality of crop are the last but one subcategory for the seed group. Precise identification and sorting of crop quality features will surge the price of the product and cut the waste. A new method, using SVM as a model, was introduced and developed by the authors to detect and identify botanical and non-botanical foreign substance embedded in cotton lint throughout harvesting and could produce the result with 95 percent accuracy in the study [42]. The study's aim was to improve quality while minimizing damage to fibres. One study

[43] included pear production, and more specifically, using SVM or SPA-SVM, a method was developed to identify and distinguish Korla fragrant pears into persistent-calyx or deciduous-calyx classes. The technique implemented hyperspectral reflectance imaging methods using ML. They were able to reach 93.3% of accuracy for deciduous-calyx and 96.7% of accuracy for persistent-calyx pears.

The authors of [44] provided the final research for this subcategory, in which a method was defined for determining and classifying the geographical source of rice samples. The technique was based on ML practices that were expanded to include the sampling of chemical components. Most precisely, with two distinct environmental areas in Brazil: Rio Grande do Sul and Goias, the key aim was to determine the regional roots of the rice. Tests indicated that the four most critical chemical components for sample classification are Rb, Cd, K and Mg. They were able to achieve 93.83 percent accuracy with the aid of ensemble learning and the random forest.

6) *Recognition of Species:*
One more area that comes under crop management is the identification of plants. The main goal is to identify and mark plant species correctly, and evade the interference of human experts, as well as speed up the classification process. In [45] a technique was developed to differentiate and identify three types of legumes, namely red beans, white beans and soybean, by patterns of leaf veins. Vein anatomy conveys detailed knowledge regarding the leaf's property. Together with colour and shape, it is an excellent method for distinguishing species. The researchers were able to reach 90.2 percent for white beans, 98.3 percent red beans 98.8 percent soybean accuracies respectively with vein leaf pictures of red, white beans and soybean along with 5-layer CNN model.

[68], implemented image-based recognition, where images are collection of different parts of plants ex. Leaf, bark or stem. Every part underwent a different approach of recognition, like for flower they chose fusion of colour, shape and texture. In order to incorporate the tropical and seasonal influences on the plant appearance, system made use of metadata

containing latitude, longitude, date, time and content. The complete system after successful recognition provided the results in the form of species name, its family and genus. ImageClef dataset was used for the experiment that has 50 classes of species. They managed to attain 98% accuracy in leaf scan sub-category and 67.3% for fruit subcategory. [69] collected 10,000 images of 100 ornamental plant species from the campus of Beijing Forestry University. They passed this dataset from a 26-layered deep learning model ResNet-26 that had 8 residual building blocks. With this model they got a recognition accuracy rate of 91.78%. Researchers of [70] had developed a model for medicinal plant identification using machine learning and computer vision algorithms. Images of 24 different plants were taken into consideration for the experiment. With these images they were able to extract features like length, breadth, area, perimeter, colour, area of hull and number of vertices. Also, they have derived certain new features from them. After comparing different models like naïve bayes, K-NN, random forest classifier neural network and support vector machine, they found out that random forest gave them best results with an accuracy of 90.1% using 10-fold cross validation method. In [71] authors picked up the morphological features for identification of botanical species. For the study they worked on 40 leaves from 30 different trees and 19 varieties of shrub family. Texture, colour and shapes were the features chosen for analysis. They have used two devices in order the capture the images: mobile and scanner. And applied various machine learning models on images captured from these devices. They selected SVM, Adaboost, random forest and ANN models for the investigation. They also have figured out that applying Computer vision algorithm proved to be more effective in terms of recognition (> 93%). Except AdaBoost other algorithms performed better. Based on this they have created a software as well named Inovtaxon Plant Species Identification for the recognition purpose.

The way crops require management in the same way most farmers also face difficulty managing livestock. Mostly livestock production and animal welfare are the major area of concern to focus on. With the assistance of Machine Learning

algorithms, one can track the health and behaviour of an animal to detect the diseases early. Further, livestock production
addresses problems in the manufacturing sector, where the
key objective of ML implementations is a precise estimation
of the financial equilibriums of the producers based on man-
ufacture line monitoring.

7) *Welfare of the Animals:*
Numerous publications are listed as belonging to animal welfare. In the first article [46], a framework for classifying cattle
behaviour that is based on ML models are described, that uti-
lize data obtained by magnetometer- and three-axis-accel-
erometer collar sensors. The research sought to forecast activities like the oestrus and to understand dietary shifts of the
cattle. Bagging with tree learner helped them to achieve 96%
of accuracy on cattle movements. A framework for the automated recognition and detection of chewing habits in calves
was introduced in the subsequent article [47]. The authors
established an ML-based system that implemented data from
chewing food supplement signals, such as ryegrass and hay,
along with the pattern of the data, such as ruminations and
idling. Optical FBG sensors were used to collect data. The
chewing pattern classification was performed using a decision tree which gave 94% accurate results. In another study
[48], an ML-based automatic surveillance device was proposed for monitoring animal behaviour, including recording
animal motions through depth video cameras, to detect different animal behaviours (standing, moving, sitting, eating,
and drinking). Using two depth cameras for the 3D motion of
pigs they were able to perform animal tracking and behaviour
annotation to measure their behavioural changes. Gaussian
Mixture Models were used here for the purpose. In [73] used
the 2 D imaging system in order to detect the lying positions
of folks of pigs (sternal and lateral). Support Vector Machine
was used along with image processing algorithm. Background
subtracting technique was used in order to extract animals
from background of the images obtained from RGB cameras.
For the input dataset to be fed to SVM classifier they have calculated the convex hull, perimeter and are of each edge. These
images where then converted to binary images to improve

the detection. With this they were able to achieve 94.4% accuracy in the classification and 94% for the detection phases using 2-D images. [74] used the combination of inter-frame and background-subtraction to study the actions of dairy calf. They monitored every motion act like entering an area, leaving an area, where they are stationary, moving around, everything detected with success rate of 93-97%. In future this can be the base for monitoring the welfare and health of calves on dairy fields.

8) *Livestock Production:*

Livestock production refers to studies designed to reliably forecast and approximate farming parameters to maximize the production system's economic performance. Work on [49] implemented a technique for predicting the pattern of rumen fermentation from milk fatty acids. The core objective of the analysis was to obtain the most precise estimation of rumen fermentation, which plays a vital role in evaluating intakes for milk production. Moreover, this study has shown that fatty acids of milk have perfect features to estimate the volatile fatty acid molar amounts in the rumen. The next study [50] dealt with hen production. They used six attributes and were using data collected over a seven-year period. Specifically, the SVM-based framework was designed that gave them 98 per cent accuracy for initial recognition and notification of difficulties in the profitable eggs production. A system for exact calculation of bovine weight trajectories throughout the period was proposed using SVM models in [51]. Precise measurement of cattle weights is of considerable value to breeders. The [52] deals with the creation of a method focused on zoometric estimation features and SVR models for predicting cadaver weight for beef cattle breed of Asturiana de Los Valles. The findings indicate that the system described has the ability to estimate carcass weights even before 150 days to the butchery day. The writers of [53] presented a system on convolutional neural networks (CNNs) that is used for recognition of pig faces in digital images. For training purposes, a total of 1533 colour photographs of the pig 's face were used. The key goal of the investigation was to identify animals without the necessity for radio frequency identifica-

tion (RFID) tags, which comprise the animal's distressing behaviour, are imitated across their range and time-consuming,
but they managed to achieve 96.7 per cent accuracy with the
model.

9) *Water Management:*
Water management in agriculture needs tremendous work
and plays a major role in the hydrological, climatic, and agro-
nomic equilibrium. Its right amount helps in every step of the
growth cycle of a plant. Any extra would ruin the complete
crop. If it is managed properly then it could be useful in days
when there is a shortage of water. This section consists of four
reports, produced primarily for daily, weekly, or monthly estimates of evapotranspiration. Precise evapotranspiration
measurement is a complicated procedure and is of great importance for energy management in the path of crop production, along with the operation and design of the irrigation systems. In one of the analyses in [54] the authors worked out a
numerical method for calculating the mean monthly evapotranspiration for the semi-arid and the arid regions. For estimating evapotranspiration they used highest, lowest, and
mean temperature, solar radiation, wind speed and relative
humidity as characteristics and regression models. It utilized
mean monthly temperature data from 44 meteorological stations during years 1951–2010. In one of the researches that
is devoted to ML applications in water management [55], two
models have been proposed for estimating the average evapotranspiration from temperature data obtained over the long
duration from six meteorological stations in an area (i.e.,
1961–2014). They collected data for minimum and maximum
temperature, air temperature at height of 2m, wind speed at
10m height, mean relative humidity, and duration of sun-
shine. An extreme learning machine was used in the first
model which was trained and evaluated from each station using local data. For the second approach they used a generalized regression neural network that was trained from all the
stations using the pooled data. Finally, in the research [56],
the authors established a methodology using the ELM model
that is fed with temperature data to estimate evapotranspiration weekly, for 2 meteorological stations. They aggregated

data as highest and lowest air temperature, extrinsic evapo-transpiration, and extra-terrestrial radiation. They utilized an extreme learning machine (ELM) model for the estimation. The objective was to accurately calculate the weekly evapo-transpiration for crop water management in India's arid regions based on a restricted data scenario. Temperature at daily dewpoints is also an important factor in determining predicted weather trends, as well as in measuring evapotranspiration and evaporation. The study [57] presented a methodology using machine learning, that forecasted the daily dew point temperature. The forecast information was obtained from two separate weather stations. The data included average air temperature, vapor pressure, atmospheric pressure, horizontal global solar radiation and relative humidity. Used an extreme learning machine to work on the collected data.

10) Soil Management:
This final section deals with machine learning applications on digging about properties of soil, such as the state, temperature, and moisture content. As a heterogeneous resource it requires multiple and complex processes that are difficult to understand in the research. The soil properties allow researchers to consider environmental processes and the effect on agriculture. A thorough evaluation of conditions of soil can help to manage soil effectively. Soil temperature has a substantial role in the comprehensive study of a region's climate change effects and ecological conditions. It is an important climatic parameter that regulates the mechanisms of interactions amongst soil and atmosphere. In fact, soil humidity plays a significant part in crop production variability. Nevertheless, soil estimates are typically laborious and extravagant, so the use of machine learning techniques based statistical analysis will achieve a cheap and efficient resolution for accurate soil estimation. The first analysis is the dissertation of [58]. More precisely, a method for evaluating soil drying for agricultural planning was provided in this report. With data on evapotranspiration and precipitation, the device accurately assesses soil drying in a region located in U.S Urbana, IL. K-nearest neighbour and backpropagation algorithm were adopted to perform the evaluation using precipitation and potential

evapotranspiration data and reached up to 91-94% of accuracy. The purpose of this approach was to help taking remote decisions in the management of agriculture. The second analysis [59] is for soil prediction. The study compares the four components moisture content (MC), total nitrogen (TN), and soil organic carbon (OC) prediction using regression models (LS-SVM/SVM, and regression / Cubist). Precisely, the researchers used a visible-near infrared (VIS-NIR) spectrophotometer to obtain soil spectra from 140 wet samples and unprocessed from the upper soil forms layer, namely Luvisol. The soil samples were gathered in the month August 2013 after harvesting of (wheat) crops from a cultivable field in Premslin, Germany. They inferred that soil characteristics can be precisely predicted to improve soil management. In a 3rd study[60], the authors discovered a new process focused on a self-adaptive evolutionary-extreme learning machine (SAE-ELM) model and regular weather data (high, low and average air temperature, atmospheric pressure, global solar radiation) to measure soil temperature at six diverse elevations of 5, 10, 20, 30, 50 , and 100 cm in two unlike areas of Iran's environmental conditions; Kerman and Bandar Abbas. The objective was to accurately measure soil temperature. An innovative approach for estimating soil moisture was presented in the last study [61], based on ANN models (Multi-layer perceptron (MLP) and radial basis function (RBF)) and used data of pressures or forces applied on a no-till chisel opener using sensors.

The below figure summarizes what all algorithms can be applied in which area of agriculture based on above discussion:

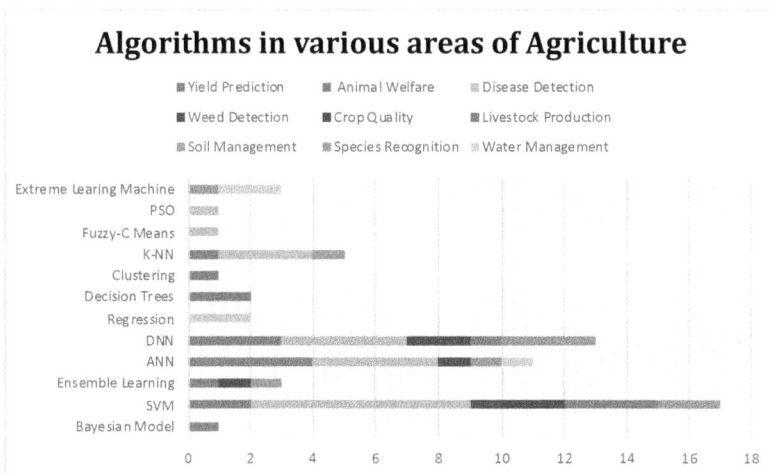

Figure 5. Summary of majorly used algorithms.

Proposed Suggestions

Since there is automation coming in every sector then how come agricultural sector can be left behind. There are various open areas where automation can be implemented in reality. Starting with the irrigation system, where automation can help in achieving the optimisation in terms of water usage. This is driven by the soil moisture sensor which would be monitoring the level of moisture in the soil and take necessary decision regarding watering of farm. Deep Learning techniques can be used in crop yield prediction, disease detection, plant recognition and fruit counting as all these involves processing of images and extracting of in-depth features. Using of pre-trained models along with the deep learning models aids in improving the accuracy and most importantly reduces the computational cost, as there is no need to calculate the weights of millions of parameters. In order to get rid of unwanted plants(weed) models like CNN, RNN and other state-of-the-art pieces are proven more effective.

Future Scope and Challenges

Various fields are left unexplored. Agriculture gives us a great opportunity to formulate new problem statements from the available datasets and the current working strategies. Few areas where the algorithms can be applied are, if collect data for rain patterns, irrigation system, crops, and fertilizer use as a time series pattern we can deduce that whether there is an increase or decrease in crop yield due to the usage of some fertilizers, we can also have k-means clustering applied on the same to identify which one gives best results. Also, it can help suggest crops based on past rainfall patterns, so farmers would be well prepared before the monsoon. We can try applying different algorithms on the currently available dataset and try to improve the accuracies of the model by pre-processing the data, augmenting the data to increase the size of the dataset. We can try to make use of satellite images to estimate the crop yield, detect weed, and track the animal movements. With the help of data regarding the consumption pattern of people in a particular village/city/state/country farmer could be aware of the amount of crop production they should have. Also, machine learning might help farmers to decide when is the best time or best price for their crops to bring in the market. The way we have been applying machine learning to predict cancer and other medical conditions of humans' same way can be helpful for animals and livestock. These technologies can also help to optimize and analyse the soil and water resources. With everyday new research in the field of computer vision, there are now various new models for image detection. These models like CNN, RCNN AlexNet, ResNet, GoogleNet, etc would prove to be a very useful tool in the plant disease detection and weed detection area. Deep learning methods might be used to detect the type of flowers and plant and would help farmers to provide a sustainable environment to grow. With the help of IoT, farmers could be able to generate a greenhouse environment for farming and can monitor everything remotely. Also, through sensors water level can be monitored for irrigation purposes. But most of the systems are still on paper, various automations are also done but they seem to be very costly in the current scenario. If these things go cheaper, farmers would confidently try to incorporate the strategies in their current methods. Another option could be to have centrally monitoring devices operated by some skilled worker and that can help farmers of specific areas. This way it's a win-win situation for everyone. If the IOT devices are deployed in every field it would be lot easier to extract and track real world data. Every farmer or an organisation can consume the data through various devices like desktop, laptops or mobile phone, so they are well versed with what their field actually needed. AI enabled, human like assistance would be very much helpful to farmers in terms of getting the solutions to there problems onsite without waiting for the government to take actions. With future predictions of crop yields, weather forecasting, weed detection, fertilizer requirements farmers would be well prepared for any new challenges that could come.

Conclusion

There are many areas where applying machine learning techniques is still a challenging task. But there is always a good scope to improve the existing models. Many new algorithms evolved from the existing theory of models. These models can be used in place of an older one to enhance the accuracy and performance of the system. Few models are been trained on synthetic data due to a lack of actual data. So, this is also an open task of collecting the original data in the form of images, surveys, and tests. Also, there various available techniques of pre-processing the current data and the creation of new data from existing data. These can be used to make our system more robust. Agricultural management systems are developing or evolving into actual artificial intelligence systems by embracing machine learning for sensor data, providing more sensible recommendations and visions for successive conclusions and actions with the eventual potential for performance improvement. The combination of automated data collection, data processing and analysis, machine learning implementation, and decision-making or guidance would include realistic tolls that are compatible with so-called knowledge-based farming to improve production rates and the quality of bio-products.

References
1) Jha, K., Doshi, A., Patel, P. and Shah, M., 2019. A comprehensive review on automation in agriculture using artificial intelligence. Artificial Intelligence in Agriculture, 2, pp.1-12.
2) Liakos, K., Busato, P., Moshou, D., Pearson, S. and Bochtis, D., 2018. Machine Learning in Agriculture: A Review. Sensors, 18(8), p.2674.
3) Tom Mitchell in his book Machine Learning
4) U. Shruthi, V. Nagaveni and B. K. Raghavendra, "A Review on Machine Learning Classification Techniques for Plant Disease Detection," 2019 5th International Conference on Advanced Computing & Communication Systems (ICACCS), Coimbatore, India, 2019, pp. 281-284, doi: 10.1109/ICACCS.2019.8728415.

5) Diptesh Majumdar, Arya Ghosh, Dipak Kumar Kole, Aruna Chakraborty, and Dwijesh Dutta Majumder, "Application of Fuzzy C-Means Clustering Method to Classify Wheat Leaf Images based on the presence of rust disease", Proceedings of the 3rd International Conference on Frontiers of Intelligent Computing: Theory and Applications, Vol. 327, 2015, pp. 277-284.

6) Kiran R. Gavhale, Ujwalla Gawande and Kamal O. Hajari, "Unhealthy region of citrus leaf detection using image processing techniques", IEEE International Conference on Convergence of Technology (I2CT), Pune 2014, pp. 1-6.

7) P. B. Padol and A. A. Yadav, "SVM classifier based grape leaf disease detection," 2016 Conference on Advances in Signal Processing (CASP), Pune, 2016, pp. 175-179, doi: 10.1109/CASP.2016.7746160.

8) A. N. I. Masazhar and M. M. Kamal, "Digital image processing technique for palm oil leaf disease detection using multiclass SVM classifier," 2017 IEEE 4th International Conference on Smart Instrumentation, Measurement and Application (ICSIMA), Putrajaya, 2017, pp. 1-6, doi: 10.1109/ICSIMA.2017.8311978.

9) M. Islam, Anh Dinh, K. Wahid and P. Bhowmik, "Detection of potato diseases using image segmentation and multiclass support vector machine," 2017 IEEE 30th Canadian Conference on Electrical and Computer Engineering (CCECE), Windsor, ON, 2017, pp. 1-4, doi: 10.1109/CCECE.2017.7946594.

10) Nithesh Agarwal, Jyothi Singhai and Dheeraj K. Agarwal, "Grape Leaf Disease Detection and Classification Using Multi-Class Support Vector Machine", proceeding of IEEE International Conference on Recent Innovations in Signal Processing and Embedded Systems (RISE), Bhopal 2017, pp. 238-244.

11) S. Hossain, R. M. Mou, M. M. Hasan, S. Chakraborty and M. A. Razzak, "Recognition and detection of tea leaf's diseases using support vector machine," 2018 IEEE 14th International Colloquium on Signal Processing & Its Applications (CSPA), Batu Feringghi, 2018, pp. 150-154, doi: 10.1109/CSPA.2018.8368703.

12) S. Kaur, S. Pandey and S. Goel, "Semi-automatic leaf disease detection and classification system for soybean culture", IET Image Processing, vol. 12, no. 6, pp. 1038-1048, 2018. Available: 10.1049/iet-ipr.2017.0822.

13) D. Al Bashish, M. Braik and S. Bani-Ahmad, "A framework for detection and classification of plant leaf and stem diseases," 2010 International Conference on Signal and Image Processing, Chennai, 2010, pp. 113-118, doi: 10.1109/ICSIP.2010.5697452.

14) Keyvan Asefpour Vakilian and Jafar Massah, "An artificial neural network approach to identify fungal diseases of cucumber (Cucumis

sativus L.) Plants using digital image processing", Archives of Phytopathology and Plant Protection, Vol. 46, Issue 13, Taylor & Francis 2013, pp. 1580-1588.

15) M. Dhakate and Ingole A. B., "Diagnosis of pomegranate plant diseases using neural network," 2015 Fifth National Conference on Computer Vision, Pattern Recognition, Image Processing and Graphics (NCVPRIPG), Patna, 2015, pp. 1-4, doi: 10.1109/NCVPRIPG.2015.7490056.

16) Ramakrishnan M. and Sahaya Anselin Nisha A., "Groundnut leaf disease detection and classification by using back propagation algorithm," 2015 International Conference on Communications and Signal Processing (ICCSP), Melmaruvathur, 2015, pp. 0964-0968, doi: 10.1109/ICCSP.2015.7322641.

17) Srdjan Sladojevic, Marko Arsenovic, Andras Anderla, Dubravko Culibrk, and Darko Stefanovic, "Deep Neural Networks Based Recognition of Plant Diseases by Leaf Image Classification", Computational Intelligence and Neuroscience, Article ID 3289801, 2016.

18) Sharada P. Mohanty, David P. Hughes, and Marcel Salathe, "Using Deep Learning for Image-Based Plant Disease Detection", Frontiers in Plant Science, Vol. 7, Article 1419, 2016.

19) Serawork Wallelign, Mihai Polceanu, and Cedric Buche, "Soybean Plant Disease Identification Using Convolutional Neural Network", International Florida Artificial Intelligence Research Society Conference (FLAIRS-31), Melbourne, United States 2018, pp. 146- 151.

20) Konstantinos P. Ferentinos, "Deep learning models for plant disease detection and Diagnosis", Computers and Electronics in Agriculture, Vol. 145, Elsevier 2018, pp. 311-318.

21) Umapathy Eaganathan, Jothi Sophia, Vinukumar Lackose, Feroze Jacob Benjamin, "Identification of Sugarcane Leaf Scorch Disease using K-means Clustering Segmentation and KNN based Classification", International Journal of Advances in Computer Science and Technology (IJACST), Vol. 3, No. 12, Special Issue of ICCEeT, Dubai, 2014, pp. 11- 16.

22) A. Parikh, M. S. Raval, C. Parmar and S. Chaudhary, "Disease Detection and Severity Estimation in Cotton Plant from Unconstrained Images," 2016 IEEE International Conference on Data Science and Advanced Analytics (DSAA), Montreal, QC, 2016, pp. 594-601, doi: 10.1109/DSAA.2016.81.

23) Gautham Kaushal, Rajini Bala, "GLCM and KNN based Algorithm for Plant Disease Detection", International Journal of Advanced Research in Electrical, Electronics and Instrumentation Engineering, Vol. 6, Issue 7, 2017, pp. 5845-5852.

24) S. Ghosh and S. Koley, "Machine learning for soil fertility and plant nutrient management using backpropagation neural networks," International Journal on Recent and Innovation Trends in Computing

and Communication ISSN: 2321-8169, Vol. 2, no. 2, pp. 292 – 297, 2010.

25) W. Okori and J. Obua, "Machine learning classification technique for famine prediction," In Proceedings of the world congress on engineering, vol. 2, pp. 991-996, 2011.

26) Omolola M. Adisa, Joel O. Botai, Abiodun M. Adeola, Abubeker Hassen, Christina M. Botai, Daniel Darkey and Eyob Tesfamariam, "Application of Artificial Neural Network for Predicting Maize Production in South Africa", Sustainability,2019,11,1145

27) Niketa Gandhi, Owaiz Petkar, Leisa J. Armstrong." Rice crop yield prediction using artificial neural networks", DOI: 10.1109/TIAR.2016.7801222

28) K. A. Shastry, H. Sanjay, and A. Deshmukh, −A Parameter Based Customized Artificial Neural Network Model for Crop Yield Prediction," Journal of Artificial Intelligence, vol. 9, no. 1, pp. 23–32, Jan. 2016

29) Efroymson, M.A. Multiple regression analysis. Math. Methods Digit. Comput. 1960, 1, 191–203.

30) Craven, B.D.; Islam, S.M.N. Ordinary least-squares regression. SAGE Dict. Quant. Manag. Res. 2011, 224–228.

31) Friedman, J.H. Multivariate Adaptive Regression Splines. Ann. Stat. 1991, 19, 1–67.

32) Quinlan, J.R. Learning with continuous classes. Mach. Learn. 1992, 92, 343–348.

33) Cleveland, W.S. Robust locally weighted regression and smoothing scatterplots. J. Am. Stat. Assoc. 1979, 74,829–836.

34) Goodfellow, I.; Bengio, Y.; Courville, A. Deep Learning; MIT Press: Cambridge, MA, USA, 2016; pp. 216–261.

35) Smola, A. Regression Estimation with Support Vector Learning Machines. Master's Thesis, The Technical University of Munich, Munich, Germany, 1996; pp. 1–78.

36) Suykens, J.A.K.; Van Gestel, T.; De Brabanter, J.; De Moor, B.; Vandewalle, J. Least Squares Support Vector Machines; World Scientific: Singapore, 2002; ISBN 9812381511.

37) Galvão, R.K.H.; Araújo, M.C.U.; Fragoso, W.D.; Silva, E.C.; José, G.E.; Soares, S.F.C.; Paiva, H.M. A variable elimination method to improve the parsimony of MLR models using the successive projections algorithm. Chemom. Intell. Lab. Syst. 2008, 92, 83–91.

38) Ensembling Learning Models: //blog.statsbot.co/ensemble-learning-d1dcd548e936.

39) Pantazi, X.E.; Tamouridou, A.A.; Alexandridis, T.K.; Lagopodi, A.L.; Kashefi, J.; Moshou, D. Evaluation of hierarchical self-organizing maps for weed mapping using UAS multispectral imagery. Comput. Electron. Agric.2017, 139, 224–230.

40) Pantazi, X.-E.; Moshou, D.; Bravo, C. Active learning system for weed species recognition based on hyperspectral sensing. Biosyst. Eng. 2016, 146, 193–202.

41) Binch, A.; Fox, C.W. Controlled comparison of machine vision algorithms for Rumex and Urtica detection in grassland. Comput. Electron. Agric. 2017, 140, 123–138.

42) Zhang, M.; Li, C.; Yang, F. Classification of foreign matter embedded inside cotton lint using short wave infrared (SWIR) hyperspectral transmittance imaging. Comput. Electron. Agric. 2017, 139, 75–90.

43) Hu, H.; Pan, L.; Sun, K.; Tu, S.; Sun, Y.; Wei, Y.; Tu, K. Differentiation of deciduous-calyx and persistent-calyx pears using hyperspectral reflectance imaging and multivariate analysis. Comput. Electron. Agric. 2017, 137,150–156.

44) Maione, C.; Batista, B.L.; Campiglia, A.D.; Barbosa, F.; Barbosa, R.M. Classification of the geographic origin of rice by data mining and inductively coupled plasma mass spectrometry. Comput. Electron. Agric. 2016, 121, 101–107.

45) Grinblat, G.L.; Uzal, L.C.; Larese, M.G.; Granitto, P.M. Deep learning for plant identification using vein morphological patterns. Comput. Electron. Agric. 2016, 127, 418–424.

46) Dutta, R.; Smith, D.; Rawnsley, R.; Bishop-Hurley, G.; Hills, J.; Timms, G.; Henry, D. Dynamic cattle behavioral classification using supervised ensemble classifiers. Comput. Electron. Agric. 2015, 111, 18–28.

47) Pegorini, V.; Karam, L.Z.; Pitta, C.S.R.; Cardoso, R.; da Silva, J.C.C.; Kalinowski, H.J.; Ribeiro, R.; Bertotti, F.L.; Assmann, T.S. In vivo pattern classification of ingestive behavior in ruminants using FBG sensors and machine learning. Sensors 2015, 15, 28456–28471

48) Matthews, S.G.; Miller, A.L.; Plötz, T.; Kyriazakis, I. Automated tracking to measure behavioral changes in pigs for health and welfare monitoring. Sci. Rep. 2017, 7, 17582.

49) Craninx, M.; Fievez, V.; Vlaeminck, B.; De Baets, B. Artificial neural network models of the rumen fermentation pattern in dairy cattle. Comput. Electron. Agric. 2008, 60, 226–238

50) Morales, I.R.; Cebrián, D.R.; Fernandez-Blanco, E.; Sierra, A.P. Early warning in egg production curves from commercial hens: A SVM approach. Comput. Electron. Agric. 2016, 121, 169–179.

51) Alonso, J.; Villa, A.; Bahamonde, A. Improved estimation of bovine weight trajectories using Support Vector Machine Classification. Comput. Electron. Agric. 2015, 110, 36–41.

52) Alonso, J.; Castañón, Á.R.; Bahamonde, A. Support Vector Regression to predict carcass weight in beef cattle in advance of the slaughter. Comput. Electron. Agric. 2013, 91, 116–120.

53) Hansen, M.F.; Smith, M.L.; Smith, L.N.; Salter, M.G.; Baxter, E.M.; Farish, M.; Grieve, B. Towards on-farm pig face recognition using convolutional neural networks. Comput. Ind. 2018, 98, 145–152.

54) Mehdizadeh, S.; Behmanesh, J.; Khalili, K. Using MARS, SVM, GEP, and empirical equations for estimation of monthly mean reference evapotranspiration. Comput. Electron. Agric. 2017, 139, 103–114.

55) Feng, Y.; Peng, Y.; Cui, N.; Gong, D.; Zhang, K. Modeling reference evapotranspiration using extreme learning machine and generalized regression neural network only with temperature data. Comput. Electron. Agric. 2017, 136, 71–78.

56) Patil, A.P.; Deka, P.C. An extreme learning machine approach for modeling evapotranspiration using extrinsic inputs. Comput. Electron. Agric. 2016, 121, 385–392.

57) Mohammadi, K.; Shamshirband, S.; Motamedi, S.; Petkovi´c, D.; Hashim, R.; Gocic, M. Extreme learning machine based prediction of daily dew point temperature. Comput. Electron. Agric. 2015, 117, 214–225.

58) Coopersmith, E.J.; Minsker, B.S.; Wenzel, C.E.; Gilmore, B.J. Machine learning assessments of soil drying for agricultural planning. Comput. Electron. Agric. 2014, 104, 93–104.

59) Morellos, A.; Pantazi, X.-E.; Moshou, D.; Alexandridis, T.; Whetton, R.; Tziotzios, G.; Wiebensohn, J.; Bill, R.; Mouazen, A.M. Machine learning-based prediction of soil total nitrogen, organic carbon and moisture content by using VIS-NIR spectroscopy. Biosyst. Eng. 2016, 152, 104–116.

60) Nahvi, B.; Habibi, J.; Mohammadi, K.; Shamshirband, S.; Al Razgan, O.S. Using a self-adaptive evolutionary algorithm to improve the performance of an extreme learning machine for estimating soil temperature. Comput. Electron. Agric. 2016, 124, 150–160.

61) Johann, A.L.; de Araújo, A.G.; Delalibera, H.C.; Hirakawa, A.R. Soil moisture modeling based on the stochastic behavior of forces on a no-till chisel opener. Comput. Electron. Agric. 2016, 121, 420–428.

62) Wei, M.C.F.; Maldaner, L.F.; Ottoni, P.M.N.; Molin, J.P. Carrot Yield Mapping: A Precision Agriculture Approach Based on Machine Learning. AI 2020, 1, 229-241.

63) Basavaraj S. Anami, Naveen N. Malvade, Surendra Palaiah, Deep learning approach for recognition and classification of yield affecting paddy crop stresses using field images, Artificial Intelligence in Agriculture, Volume 4,2020,Pages 12-20,ISSN 2589-7217.

64) Vijai Singh,Sunflower leaf diseases detection using image segmentation based on particle swarm optimization, Artificial Intelligence in Agriculture, Volume 3,2019,Pages 62-68,ISSN 2589-7217

65) Alhnaity, Bashar & Pearson, Simon & Leontidis, Georgios & Kollias, Stefanos. (2019). Using Deep Learning to Predict Plant Growth and Yield in Greenhouse Environments.

66) Yu J, Schumann AW, Cao Z, Sharpe SM and Boyd NS (2019) Weed Detection in Perennial Ryegrass with Deep Learning Convolutional Neural Network. Front. Plant Sci. 10:1422.

67) Muhammad Hamza Asad, Abdul Bais,Weed detection in canola fields using maximum likelihood classification and deep convolutional neural network,Information Processing in Agriculture,Volume 7, Issue 4,2020,Pages 535-545,ISSN 2214-3173

68) S. Purohit, R. Viroja, S. Gandhi and N. Chaudhary, "Automatic plant species recognition technique using machine learning approaches," 2015 International Conference on Computing and Network Communications (CoCoNet), Trivandrum, 2015, pp. 710-719

69) Y. Sun, Y. Liu, G. Wang, H. Wang, Deep learning for plant identification in natural environment. Comput. Intell. Neurosci. (2017)

70) Begue, Adams & Kowlessur, Venitha & Singh, Upasana & Mahoodally, Fawzi & Pudaruth, Sameerchand. (2017). Automatic Recognition of Medicinal Plants using Machine Learning Techniques. International Journal of Advanced Computer Science and Applications. 8. 10.14569/IJACSA.2017.080424.

71) Bambil, D., Pistori, H., Bao, F. et al. Plant species identification using color learning resources, shape, texture, through machine learning and artificial neural networks. Environ Syst Decis 40, 480–484 (2020).

72) Pearline, Anubha & Kumar, Sathiesh & Harini, S. (2019). A study on plant recognition using conventional image processing and deep learning approaches. Journal of Intelligent & Fuzzy Systems. 36. 1-8. 10.3233/JIFS-169911.

73) Abozar Nasirahmadi, Barbara Sturm, Anne-Charlotte Olsson, Knut-Håkan Jeppsson, Simone Müller, Sandra Edwards, Oliver Hensel, Automatic scoring of lateral and sternal lying posture in grouped pigs using image processing and Support Vector Machine, Computers and Electronics in Agriculture, Volume 156, 2019, Pages 475-481, ISSN 0168-1699.

74) Guo, Y.; He, D.; Chai, L. A Machine Vision-Based Method for Monitoring Scene-Interactive Behaviors of Dairy Calf. Animals 2020, *10*, 190.

AI application in Agriculture Sector

Manisha Verma
Computer Science & Engineering Department, Hindustan College
of Science & Technology, Mathura, Uttar Pradesh, India

Abstract: The use of Artificial Intelligence (AI) is constantly increasing in the recent past. AI is being used in many different sectors like industrial sector, healthcare sector, etc. due to the advantages it provides. In agricultural sector also, AI is finding and making its place. Agriculture is a sector which is undoubtedly one of the most important and necessary sectors. The application of AI in agriculture can be of great help. This book chapter presents a systematic study of AI application in agriculture sector based on literature. A real world AI application example in agriculture is also discussed. Along with it, the advantages of AI in agriculture are also discussed in this chapter.

1. Introduction

Agriculture is crucial for everyone. World population depends on agriculture largely for food. With growth in world population, the food requirement also grows accordingly. In such scenarios, increasing the food production is necessary.

Hence, any techniques, methods, technologies, etc. which can help in agriculture, which can make the work of the farmers easier could be beneficial and useful. For example, AI application in agriculture could be beneficial and useful in agriculture. This is discussed later in this chapter.

Agriculture is largely responsible for food production and along with it, agriculture is a very important contributor towards economy. It is very useful for the economy as well. Different counties on a regular basis import and export varieties of fruits, vegetables, etc. such agriculture produce throughout the world.

There are so many different types of agricultural produce across the world. Let's discuss some of them. The agricultural produce forms staple food of people like wheat, rice, vegetable, fruits, etc. to things like cotton which is not consumable but is having a lot many important uses.

Cotton is used in many places, like, for example, it is used in health sector, clothing sector, furnishing sector and various other sector. In health sector, it is having a lot of importance. Cotton is largely used in the health sector.

Similarly in the clothing sector, furnishing sector and at various other places cotton is used largely. The grown raw form of cotton is processed as per applicability and then used as per requirement.

There are many varieties of agricultural produce. Consider vegetable for example, then it is noticed that there so many types of vegetables grown across the world. Some of them are pumpkin, tomato, onion, cabbage, potato, etc.

Even when it comes to fruits, various types of fruits are grown across the world such as mango, watermelon, banana, guava, apple, cherry, etc. and a large list of dry fruits are also grown across various places in the world such as walnut, almond, cashew, apricot, dry figs, etc.

All the different type of agricultural produce which a person includes in their daily diet that becomes a part of their staple food, provides them with all the needed nutrients, vitamins, protein, etc. essential requirements which are much needed by individuals and they cannot keep themselves void of these essential requirements.

Whenever there is less agriculture produce in different places across the world, it has serious consequences for the world. Especially when the places which export large quantities of agricultural produce like rice, wheat, etc. that are part of the staple food of many people across the world, have less agriculture produce, the consequences for the world become worse.

The consequences include shortage of food which is and shortage of food is a great cause of concern. It has many ill effects on people. Like for example, at many places, food becomes completely unavailable and unavailability of staple food or part of staple food can cause mal-nourishment and other deficiencies among people. Where the food is still available in less quantities, it leads to inflation in prices of food.

To add to this, the worst case scenario is that each year the world witnesses people losing their lives due to hunger. All these things should not take place and in order to ensure that all this does not take place, the agriculture production level should never be allowed to come down.

Finding of new ways to increase the agricultural production can be effective. Along with finding ways to help increase the agricultural production, ways to help the farmers in their agricultural work can be beneficial and useful. Also, food wastage needs to be avoided completely.

2. AI in Agriculture

AI means machines or any system which copies and duplicates intelligence like that of humans. AI can help largely in different sectors like education, health, industrial, etc. including agriculture.

Agriculture involving AI can impact agriculture in a way that is of advantage for agricultural sector. In the recent past, use of AI in agriculture is coming in limelight. Different companies have also started working in this direction.

An example of a company using AI in agriculture is also discussed later in the chapter. Since agriculture is crucial for everyone and AI can help largely in agriculture, including it would be of help for agriculture.

3. AI Applications in Agriculture

AI can have various applications in agriculture. In this section, we will be discussing the following application of AI in agriculture.

- Rainfall forecasting using AI
- AI for estimation of soil moisture
- AI for modeling hydroponic plant growth dynamic response to root zone temperature

- AI for identification and classification of diseases from paddy plant images

3.1 Rainfall Forecasting using AI

With weather conditions changing across the world, it becomes difficult for farmers to plan out sowing of seeds for crops to grow, to plan out irrigation, etc. other related work in agriculture. With the help of artificial intelligence, farmers can forecast rainfall to analyze weather conditions to help them plan out different things and work accordingly.

In the article [25], the authors have explained a study which is comparison based study. Comparison of three different neural network architectures is done to forecast rainfall in Thanjavur district of southern province Tamil Nadu, India. The neural network's considered for comparison are GRNN which stands for generalized regression neural network, RBNN which stands for radial basis function neural network and BPNN which stands for back propagation neural network [25].

Utilizing training data set, the different models were trained and for development of the model, MATLAB was used. After which, to know the models accuracy, they were tested for it based on the test data which was available and it was found the optimum result for prediction is given by RBNN [25].

3.2 AI for Estimation of Soil Moisture

In this particular paper [8], the authors have tried to find out how well do artificial neural networks (ANNs) work for getting an estimate of the content of soil moisture. To find out this, different datasets from different soil covers were taken. So datasets from three experimental soil covers (D1, D2, and D3), having a thickness of 0.50 m, 0.35m, and 1.0 m, consisting of a peat mineral mix thin layer over varying thickness of till, are taken into consideration in this paper [8]. At till layers and peat layers, volumetric soil moisture contents were modeled as a function of air temperature, precipitation, net radiation, and ground temperature at different layers. Initially it was found that ground temperature is the state variable which is having major influence for soil moisture characterization [8].

So with the aim of obtaining as much as possible information from ground temperature, a higher-order neural networks (HONNs) model was developed in order to characterize dynamics of soil moisture [8].

As compared to traditional artificial neural network, for some simulations of soil moisture, the HONNs resulted in relatively higher correlation coefficient. In order to make the performance of the model better and get best results, inputs which were time-lagged were used [8].

As compared to a concept based model which was developed earlier for the purpose of predicting the depth-averaged soil moisture content, ANN models performance was better. It is noticed that with ANN, modeling of soil moisture is filled with challenges but it can be achieved [8].

The performance of ANN depends on results from the study indicate that modeling of soil moisture using ANNs is challenging but achievable, and the structure and formation of soil covers to a large extent influence the performance of ANNs [8].

3.3 AI for Modeling Hydroponic Plant Growth Dynamic Response to Root Zone Temperature

In hydroponic system, knowing the favorable temperature of root zone while cultivation of it is going on, could allow to have betterment in growth of the plant. A favorable control strategy can be decided by identification of eco-physiological process with the help of a dynamic model [12].

Since the eco-physiological processes of plants are complex, it becomes hard for developing out a dynamic model of plant growth responses to root zone temperature. The authors in [12] have proposed and developed a dynamic model of responses of growth of plant to root zone temperature.

For developing the model, NARX neural networks i.e. non-linear autoregressive with exogenous input neural networks were used. These neural networks are a specific type or class of artificial neural networks. This model was used for plant of chili pepper. For a total of 60 days during cultivation in a growth chamber [12].

Five different datasets consisting of dynamic response of growth of plant for identification of system were obtained. The results the authors received indicate that neural network is of great use with encouraging performance for modeling hydroponic plant growth dynamic response to root zone temperature [12].

3.4 AI for Identification and Classification of Diseases from Paddy Plant Images

In this paper [24], authors want to find out blast disease. This will help in reduction in crop loss and can help in more production of rice in an effective way. Paddy production faces a threat from pest and diseases. Weed is also not good for paddy plants [24].

Hence, the authors in this paper have utilized deep convolutional neural networks and carried out image classification using it. Images are collected from Image Net dataset. The developed model is able to classify 16 classes of paddy pests, diseases and weeds [24].

The results show that we can effectively detect and recognize the rice diseases and pests including healthy plant class using a deep convolutional neural network, with the best accuracy of 96.50% [24].

4. Real-world AI Application Example in Agriculture

The soil and the nutrients that the soil contains, plays an important role in agriculture. The crop growth to a large extent depends on the soil and the nutrients it contains. It is difficult to determine the soil nutrients and its quality by just looking at it.

PEAT, which is a start-up, has developed an application which is called Plantix to help farmers. Thanks to this app, farmers can receive great help. Using artificial intelligence nutrient deficiencies in the soil and plant pests and diseases can be detected.

5. Advantages of AI in Agriculture

There are various advantages of AI in agriculture. They are as follows:

- It can help in the case of plant diseases
- AI can help in soil moisture estimation
- AI can help in rainfall prediction
- AI overall can help farmers in agricultural sector in different things

5.1 It can Help in the Case of Plant Diseases

As discussed in the chapter, AI can be of great help when it comes to identification and classification of plant diseases. In paper [24] authors have successfully made utilization of AI for the same purpose.

Also, PEAT which is a start-up, successfully makes use of AI for helping farmers. The application Plantix developed by them, help farmers indifferent ways, which includes helping in the case of plant diseases.

5.2 AI can Help in Soil Moisture Estimation

AI can help in various works carried out by farmers in the field. Helping in soil moisture estimation is one of the important advantages of AI in agriculture. AI can provide estimate of soil moisture and this is even seen in paper [8].

The soil moisture is an important parameter in understanding the irrigation requirements of plants in the field. When the soil moisture estimate is known then based on it irrigation can be carried out and this will be useful for the farmer.

5.3 AI can Help in Rainfall Prediction

Rainfall is also a very important factor in agriculture. It is necessary to look after this factor. Keeping the rainfall in mind, different types of plants are grown. Also, irrigation schedule is to be adjusted according to the rainfall.

If extra water is provided then it can harm the plants. If no water is provided then also it harm the plants. This is why according to the rainfall, irrigation has to be adjusted and rainfall being an important factor in agriculture, if it can be predicted using AI then it will be useful and helpful for the farmers.

5.4 AI Overall can Help Farmers in Agricultural Sector in Different Things

AI can help in various things related to agriculture and AI overall can help farmers. For example, AI is helpful for modeling hydroponic plant growth dynamic response to root zone temperature as seen in the chapter. This could highly benefit the plant growth and can help to have a better growth of the plant.

AI based predictive analysis could be used in agriculture. It could help to detect different issues that could occur in field and affect the plants. The solutions can be found according to it and the issues can be prevented.

6. Conclusion

AI usage has been growing in the recent past. AI is already being used in various sectors such as industrial sector, health sector, etc. This isn't surprising as the advantages of using AI come across to be helpful. In this chapter, we saw that in agricultural sector also, AI is making its place.

Agriculture is very important, and AI applications can help and benefit this sector. In this book chapter, a brief introduction regarding agriculture, followed by a discussion of using AI in agriculture is presented. A systematic study of AI application in the agriculture sector based on literature is also presented where the results stated in the literature are also discussed and presented.

From the study of the literature, it can be stated that AI can help in different work of the farmers. A real-world example of AI application in agriculture is also discussed which further indicates and shows that AI can help in different work of the farmers. The different advantages of AI in agriculture are also discussed in this chapter.

References

1. Gaitan, A.; Gaitan, V. G.; Ungurean, I. (2015) A survey on the internet of things software architecture. Int. J. Adv. Comput. Sci. Appl, 140–143.
2. Mekala, M. S.; Viswanathan, P. (2017) A novel technology for smart agriculture based on iot with cloud computing. In Proceedings of the International Conference on IoT in Social, Mobile, Analytics and Cloud (I-SMAC), pp. 75–82.
3. Markovic, D.; Koprivica, R.; Pesovic, U.; Ranic, S. (2015) Application of IoT in monitoring and controlling agricultural production. Acta Agric. Serbica, 12, 145–153.
4. Wu, S.; Austin, A.; Ivoghlian, A.; Bisht, A.; Wang, K. (2020) Long range wide area network for agricultural wireless underground sensor networks. J. Ambient Intell. Humaniz. Comput.
5. King, I. (2017) Technology: The future of agriculture. Nature, S21–S23.
6. Chen, W.-T.; Yeh, Y.-H.F.; Liu, T.-Y.; Lin, T.-T. (2016) An Automated and Continuous Plant Weight Measurement System for Plant Factory. Front. Plant. Sci., 7, 392.
7. Chan, R.W.K.; Yuen, J.K.K.; Lee, E.W.M.; Arashpour, (2015) M. Application of Nonlinear-Autoregressive-Exogenous Model to Predict the Hysteretic Behaviour of Passive Control Systems. Eng. Struct., 85, 1–10.
8. Elshorbagy, A., Parasuraman, K. (2008) On the relevance of using artificial neural networks for estimating soil moisture content. *Journal of Hydrology*, 362, pp. 1– 18.
9. Sensor based Automated Irrigation System with IOT" International Journal of Computer Science and Information Technologies, ISSN: 0975-9646, Vol. 6 (6) 2015, 5331-533
10. Foster, I.; Kesselman, C. (1998) The Grid: Blueprint for a New Computing Infrastructure; Morgan Kaufmann Publishers Inc.: San Francisco,CA, USA,.
11. Adetunji, K.E.; Joseph, M.K. (2018) Development of a cloud-based monitoring system using 4Duino: Applications in agriculture. In Proceedings of the International Conference on Advances in Big Data, Computing and Data Communication Systems, Durban, South Africa, 6–7 August; pp. 4849–4854.
12. Aji, G. K.; Hatou, K.; Morimoto, T. (2020) Modeling the Dynamic Response of Plant Growth to Root Zone Temperature in Hydroponic Chili Pepper Plant Using Neural Networks. Agriculture, 10, 234.
13. Arun Pandian, J. A., Geetharamani, B. Annette. (2019) Data Augmentation on Plant Leaf Disease Image Dataset Using Image Manipulation and Deep Learning Techniques, *IEEE 9th International Conference on Advanced Computing (IACC)*, 2019, pp. 199-204, doi: 10.1109/IACC48062.2019.8971580.
14. M. Pagliai, N. Vignozzi, S. Pellegrini, (2004) Soil structure and the effect of management practices, *Soil and Tillage Research*, Vol. 79, No. 2, pp. 131-143.
15. Liakos, K. G.; Busato, P.; Moshou, D.; Pearson, S.; Bochtis, D. (2018) Machine Learning in Agriculture: A Review: Sensors, 18, 2674.

16. Gavhale, K. R. Gawande, U., Hajari, K. O. (2014) Unhealthy region of citrus leaf detection using image processing techniques, *International Conference for Convergence for Technology-2014*, 1-6.

17. M. Pagliai, N. Vignozzi, S. Pellegrini, (2004)"Soil structure and the effect of management practices", Soil and Tillage Research, Vol. 79, No. 2, pp. 131-143.

18. G. S. Abawi, T. L. Widmer, (2000)"Impact of soil health management practices on soil borne pathogens, nematodes and root diseases of vegetable crops",Applied Soil Ecology, Vol. 15, No. 1, pp. 37-47.

19. E. M. Lopez, M. Garcia, M. Schuhmacher, J. L. Domingo, (2008)"A fuzzy expert system for soil characterization", Environment International, Vol. 34, No. 7,pp. 950-958.

20. C. Aubry, F. Papy, A. Capillon, (1998)"Modelling decision-making processes for annual crop management", Agricultural Systems, Vol. 56, No. 1, pp. 45-65.

21. R. E. Plant, (1989)"An artificial intelligence based method for scheduling crop management actions", Agricultural Systems, Vol. 31, No. 1, pp. 127- 155.

22. Boniecki, P.; Koszela, K.; Swierczy´nski, K.; Skwarcz, J.; Zaborowicz, M.; Przybył, J. (2020) Neural Visual Detection of Grain Weevil´ (Sitophilus granarius L.). Agriculture pp 10- 25.

23. Niedbała, G.; Kurasiak-Popowska, D.; Stuper-Szablewska, K.; Nawracała, J. (2020) Application of Artificial Neural Networks to Analyze the Concentration of Ferulic Acid, Deoxynivalenol, and Nivalenol in Winter Wheat Grain. Agriculture pp.10-127.

24. Bharathi, R. J. (2020) Paddy Plant Disease Identification and Classification of Image Using AlexNet Model. *The International journal of analytical and experimental modal analysis*. XII(III), pp. 1094-1098.

25. Manek, A. H. (2016) Comparative Study of Neural Network Architectures for Rainfall Prediction. *2016 IEEE International Conference on Technological Innovations in ICT For Agriculture and Rural Development (TIAR 2016).* pp. 171-174.

26. Anami, B. S, Malvade, N. N., Palaiah, S. (2020). Deep learning approach for recognition and classification of yield affecting paddy crop stresses using field images. *Artificial Intelligence in Agriculture* 4, 12-20.

27. Nasiakou, A., Vavalis, M., Zimeris, D. (2016) Smart energy for smart irrigation, *Computers and Electronics in Agriculture*, 129, pp. 74-83.

28. Pernapati, K. (2018). IoT based low cost smart irrigation system. *Second International Conference on Inventive Communication and Computational Technologies (ICICCT)*, 1312- 1315.

29. Kaur, S., Pandey, S., Goel, S. (2018). Semi-automatic leaf disease detection and classification system for soybean culture, *IET Image Processing*, 12(6), 1038-1048.

30. Purohit, S., Viroja, R., Gandhi, S., Chaudhary, N. (2015). Automatic plant species recognition technique using machine learning approaches. 2015 International Conference on Computing and Network Communications (Co-CoNet), 710-719.

31. Gutirrez, J., Francisco, J., Villa-Medina, J. F., Nieto-Garibay, A., Porta-Gándara, M. Á. (2014). Automated Irrigation System Using a Wireless Sensor Network and GPRS module, *IEEE Transactions on Instrumentation and Measurement,* 63(1), 166-176.

www.ingramcontent.com/pod-product-compliance
Lightning Source LLC
Chambersburg PA
CBHW060242230326
41458CB00094B/1407